D1338797

New City College

Please return on or before the ~~rned~~ shown below
Failure to do so will result in a fine.

WHY FONTS MATTER.

Sarah Hyndman

NEW CITY COLLEGE LIBRARY
REDBRIDGE CAMPUS

10 9 8 7 6 5 4

Virgin Books, an imprint of Ebury Publishing,
20 Vauxhall Bridge Road,
London SW1V 2SA

Virgin Books is part of the Penguin Random House group of companies
whose addresses can be found at global.penguinrandomhouse.com

 Penguin
Random House
UK

Copyright © Sarah Hyndman 2016

Sarah Hyndman has asserted her right to be identified as the author of this
Work in accordance with the Copyright, Designs and Patents Act 1988

First published by Type Tasting in 2015
This edition published by Virgin Books 2016

www.eburypublishing.co.uk

A CIP catalogue record for this book is available from the British Library

ISBN 9780753557235

Printed and bound in Italy by L.E.G.O. S.p.A.

Written and designed by Sarah Hyndman
www.typetasting.com
Twitter @TypeTasting

A special thank you to Monotype for allowing
access their extensive type library

To Tina F-P and Mum for your tenacity and inspiration

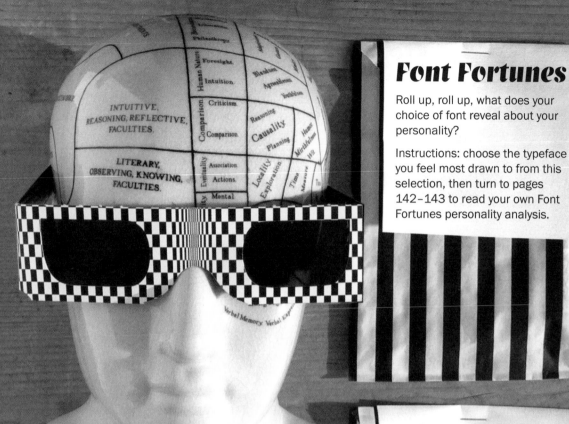

Font Fortunes

Roll up, roll up, what does your choice of font reveal about your personality?

Instructions: choose the typeface you feel most drawn to from this selection, then turn to pages 142–143 to read your own Font Fortunes personality analysis.

PHRENOLOGY
BY
L.N. FOWLER.

FONT FORTUNE

Pick Me

Miracle Font

Choice #3

Contents

Who is this book for?

Fonts are no longer just for geeks or graphic designers. Thanks to technology and the gadgets we use, more people than ever before are aware of them — from gateway fonts* like *Comic Sans* to the superstar typefaces like *Helvetica* and *Gotham*.

We are all type consumers. Typefaces/fonts play a vital role in our everyday lives. They help us to navigate, to make choices, to shop, they keep us safe and sometimes they play a game of sleight of hand.

In this book you are invited to consider your emotional response to type. Each font or typeface has a personality that influences your interpretation of the words you are reading, by evoking your emotions and setting the scene. You instinctively understand this, but it happens on a subconscious level. Becoming consciously aware of the emotional life of fonts can be entertaining and ultimately give you more control over decisions you make.

*Applying gateway drug theory to typography is to suggest that using a seemingly naive font could open the door to the use of more serious and 'hard core' typefaces.

About the author

Sarah Hyndman is a typography consultant for a range of organisations. She hosts Type Tastings, gives talks and runs workshops which range from intensive, creative workouts for designers to enlightening and entertaining sessions for non-designers.

Sarah established Type Tasting as a forum for research, skill development and exploration about the experience of type/fonts from the type consumer's point of view. Sarah does this in an entertaining and inspiring way that encourages conversation and promotes understanding beyond the design world.

Type Tasting began on Valentine's Day 2013 with an evening of 'Typographic Swearing 'n' Cussing'. Since the launch Sarah has given a TEDx talk called 'Wake Up and Smell the Fonts', and spoke about the power of type at South by Southwest (SXSW) in Austin, Texas. She has been interviewed on BBC Radio 4's *Today* programme and BBC World Service's *The Arts Hour*. Sarah has organised and curated a mass participation typography exhibition and workshops at the Victoria and Albert Museum (V&A) for the London Design Festival and teaches typography for the D&AD. She is working on collaborative research with the Crossmodal Research Laboratory at Oxford University and is currently writing her next book.

Before setting up Type Tasting Sarah freelanced as a graphic designer and then set up a design company which she ran for 10 years. Sarah gained a distinction for her Masters degree in Typo/Graphics at the London College of Communication (University of the Arts). She was subsequently invited back as a guest tutor to set up and run the year-long Experimental Typography evening course, which she did alongside her commercial practice.

www.typetasting.com
Twitter @TypeTasting

Links to the organisations mentioned can be found on page 140.

Introduction

I first fell in love with type in the 1970s as a child in a sweet shop on the way home from school. Here, unlike in the classroom where I read 'serious' typefaces, words came to life as they fizzed and popped with excitement, all shouting out their different flavours. I thought that fonts were like multi-sensory imagination grenades.

I have now been a graphic designer for almost 20 years. I began working in big agencies where I learned about the power of type and then I set up my own design company in 2003.

This book takes you through the findings of my journey of discovery, which began when I turned the tables on myself and set off to explore fonts from the type consumer's point of view.

It has been important for me to talk to as many people as possible along the way, which has involved running events, inventing games, giving talks, getting out of the graphic design world and asking lots of questions. I have read research dating back to the 1920s and, where I cannot find proven answers, I have run experiments and created my own surveys. These have been designed to gather data, but also to be fun and to get people thinking about fonts.

This curiosity has opened up new avenues that I could not have imagined, like working with scientists at Oxford University to explore how type can influence our senses.

The stories and discoveries in this book are my personal observations and are based on my experiences. As a graphic designer I am a generalist, which means I try to take a wide viewpoint and gather information from a variety of sources. As with everything else I do, this book is intended as a conversation starter.

Sarah Hyndman

1

Typefaces, they don't really matter?

Justin Webb, BBC Radio 4 *Today*

Dalston Type Safari
'Mockingbird Tapas' sign painted by Peter Hardwicke / Art Deco 'Jazz Bar' / New York-style 'Voodoo Ray's' / Blackletter 'Ta' / Fried chicken shop Flare Serif 'S' / 'Moustache Bar' icon / '70s-style *Candice* 'G' / Vernacular 'Tom's Bakery' / New York-style 'Voodoo Ray's' 'S' / Art Deco 'Rio Cinema' 'R' / 'Dalston Superstore' in neon / 'Ruby's' in condensed *Helvetica* / Mosaic lettering / 'Crown and Castle' Grotesque Sans Serif / Geometric 'Dalston Culture House' / Woodblock-inspired 'Birthday's' 'B'.

TYPEFACES, THEY DON'T REALLY MATTER?

I was sitting nervously in the BBC Radio 4 *Today* studio very early one Saturday morning trying hard not to think about the 7 million listeners who were apparently tuned in as I gave my first ever live radio interview. Time seemed to slow and my heartbeat got louder as I waited the eternity it seemed to take for the interview to begin. As the moment got closer I started to worry that not only was I losing the ability to speak, but also I could no longer remember anything—not even my name. How was I going to have a conversation about typefaces?

Finally the light saying 'live' went on and the interviewer turned to me and said, 'Typefaces, they don't really matter?'[1] This is a question that I have been asked so many times that I forgot my nerves and found myself taking on the familiar role of 'type advocate' as we chatted about the importance of type in our everyday lives.

Type consumers

We are all type consumers: we all interact with, and consume, a vast array of typefaces every day of our lives and most of the time we do this without being consciously aware of it. Type influences what we read and affects our choices because we all instinctively understand what it is communicating to us, and we have been learning to interpret the references all our lives.

As a graphic designer I actively think about type when I am working, but once I am 'off duty' I become a type consumer. At this point I generally stop paying attention to fonts. Occasionally I will notice one if it stands out for a reason; I might find it particularly pleasing, unexpected, or an inappropriate choice.

I was interested to see how many fonts I consume as I go about my everyday life, so I set myself the challenge of counting how many I encountered over the course of an hour on a Saturday morning.

Phone: check time, emails (2) / Bathroom: toothpaste, shower products, toiletries (7) / Open curtains: road markings, signs, vehicles (3) / Kitchen: coffee, fridge, clock (5) / Breakfast ingredients (7) / Toaster, microwave (3) / Living room: TV, news, advertisements (14) / Sort through books and magazines, pick a bag up off the floor (7) / Computer (5) / Items on the desk (4) / Open post (9) **Total: 67**

This takes a bit of focus as you quickly forget to notice the fonts and get distracted by the activities. Try this challenge for yourself. How many fonts do you consume in an hour?

Curating fonts

Famous brand logos can often be recognised from the type alone. What you see can be reduced down to just a fragment of the logo and you will still know instantly what the brand is. This is because your brain is a 'pattern-matching machine'. You compare what you look at to things you have seen before and catalogued in your mind. Once a logo has become familiar you no longer even need to read the words to know what they say, because you recognise it by its shape: think about the Coca-Cola™ or Google logos. You do this from an early age, as demonstrated by Anna, aged 4, whose mother Sandra tweeted 'Our daughter Anna says our car is from Boots (the chemist). What is our car?' The answer is on page 21.

We curate the logos we surround ourselves with from the range of brands we use in our everyday rituals. These include toothpaste and shampoo, the objects we select to reflect our personality, like the brands of clothes we wear and expensive lifestyle items like cars. We are 'creatures of habit' says design writer Steven Heller[2], who observes that it takes a 'diabolically brilliant' marketing plan to entice us to change our brand loyalty. But if a brand alters its typography to a style that no longer feels in keeping with our values we can be acutely aware of the change. In 2010 high street fashion shop Gap scrapped its redesign after just a few days following customer protests that the new logo looked 'cheaply, tacky, ordinary' and returned to the original logo created from the *Spire* typeface.[3]

GAP

Original design: *Spire*

Gap

Redesign: *Helvetica bold*

Luxury Bubbles Daily Bubbles Kids Bubbles

Lemon Liquid Fave Liquid Cheap Liquid

Luxury Wash Cheap Wash

Fast Wash Super Wash

Fast Clean Strong Clean Cheap Clean

Lemon Liquid Fast Liquid Cheap Liquid

Luxury Wash

Fast Wash

Luxury Hand Cream Cheap Hand Cream Smelly Hand Cream

Luxury Soap Daily Soap Kids Soap

Luxury Bubbles Kids Bubbles

Fast Clean Strong Clean Cheap Clean

Super Wash Cheap Wash

Fast Clean Strong Clean Cheap Clean

Game: Supermarket Sweep

Congratulations – you've won the lottery! To celebrate you're going to buy all the luxury items in the shop. How long will it take you to find all the luxury products?

Supermarket A

Number of luxury items:

Time taken to find them:

Now turn the page to go to Supermarket B.

Font users

After centuries of having typefaces chosen for us by designers and printers, we now live in a time where design tools are being increasingly democratised and more people than ever before are finding themselves looking down a font menu and deciding which ones to choose.

At a recent event I asked a 12-year-old for her thoughts on the typefaces I had printed out on a table for a game. I was taken aback by how articulately she explained which she would choose for different genres of books. She told me that she would change the font on her Kindle until it 'felt right' for each book she read. She preferred *Palatino* for fiction set in the past and a Sans Serif like *Futura* for her factual schoolbooks. The other youngsters in the room said they went through a similar process. When I was a child the typeface was an immutable part of a book because the designer or printer had already selected it. Since I had no choice I took typefaces for granted. Having a choice has given a generation of readers a new appreciation of type, and selecting a font has become as normal as choosing which t-shirt to wear or which cereal to have for breakfast.

Fonts save you time

Imagine a world with just one typeface. This might be an aesthetically appealing idea if you picture a 1950s shop full of packaging, with type all neatly lined up in a clean, Sans Serif style like *Helvetica*. This is the design of supermarket A (above). However, the reality is that, although this shop might look stylish and relaxing, your shopping trip would take much longer without the complex visual clues typefaces give you. You would have to stop and read the words on every pack to find out

See page 21 for the answers. It should have taken you considerably less time to find the luxury items in supermarket B where the typefaces give you visual clues.

Supermarket B

Number of luxury items:

Time taken to find them:

---

whether it is a luxury or cheap product, the full fat or healthy option, your favourite brand, the one you would never buy, or your guilty pleasure.

Fonts help you to choose

Typefaces help you to decide who you would trust to do a professional job for you. For example, which of these three lawyers would you hire? (See page 57).

(a)   (b)   (c)

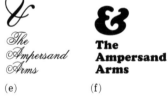

(d)   (e)   (f)

When you are in a new town the signs over the doors help you to choose where to eat and drink. Which of the three 'The Ampersand Arms' on the left do you think would suit you and why? (See page 133).

However, signs can occasionally be misleading. For example, I regularly walk past a pub with a sign in an Ornate Script similar to example (e). This is a style you are used to seeing on wine labels and wedding invitations, so you could assume that it is a wine bar or somewhere to go for a quiet, romantic drink. The reality is quite the opposite in fact, as you find yourself in a noisy, beer drinkers' pub with football on the large television screens.

Fonts keep you safe

When driving, you need to know at a glance whether a sign is official and you must react quickly, or

Clearview

Frutiger

Helvetica

Serif to Sans Serif

AB
Serif

AB

AB
Sans serif

Destination: Paris
Art Nouveau, e.g. *Arnold Boecklin*

Destination: London
Johnston is the official London Transport typeface

DESTINATION: NEW YORK
Geometric Sans Serif e.g. *Twentieth Century*

whether it is simply an advert that you can read at your leisure (or choose to ignore). Official road signs fit an established set of visual conventions. They use clear and neutral type which is highly readable, and generally in a Sans Serif style because these have proved in tests to be the most readable from a distance.

The road sign system in Britain was created in the 1960s by Jock Kinneir and Margaret Calvert[4], who designed the two new typefaces *Transport* and *Motorway* as part of the project. Life-sized signs were made and tested for legibility in an underground car park and in London's Hyde Park. Kinneir and Calvert established that words set in upper and lower case are more legible than those in just upper case, as they have silhouettes that can be distinguished from a distance. This was considered controversial at the time.

LEGIBILITY Legibility

By contrast, the advertising signs that sit alongside the road signs shouting out their messages (above) can easily be distinguished by their decorative and colourful letters, each with their own brand type style, colour and images.

Fonts show you the way...

Official directional signs in a visually cluttered and busy airport are generally easy to recognise, even when you are in unfamiliar surroundings or the signs are in a different language. According to design writer Alissa Walker[5] there are just three typefaces used by around three quarters of the airports in the world: *Clearview*, *Frutiger* and *Helvetica* (shown above). These are chosen for their clarity and legibility and are similar in style to the type found on road signs.

... and tell you where you are

Once you have arrived at your destination, you know immediately if you are in another country from the less familiar style of lettering on the signs and businesses. In Paris, the influences of Art Nouveau can still be widely seen, for example on the Métro signs. The unofficial typeface of London is *Johnston*, which is used across the London transport system. Geometric Sans Serif styles like *Twentieth Century* were popular on signage in New York in the mid 20th century and these inspired Tobias Frere-Jones's *Gotham* typeface, made famous when it was used for Obama's 2008 presidential campaign.

ILFRACOMBE IS A CURIOUS PLACE

If you were to be parachuted into an unknown location it would be the content and style of the signage that would give you some of the biggest clues to your whereabouts. You can test this out by playing the online game GeoGuessr.com,[6] which transports you to a random location somewhere in the world using Google Street View. The challenge is to see how quickly you can work out where you are.

Typographic DNA

The combinations of type found on signage reveal a great deal about a city, town or specific area. They reflect the social, economic and historical development of the area and create their own, unique typographic DNA.

I take people on Type Safaris through Dalston in East London. These are guided walks exploring the signage with a photography challenge to take part in along the way. We sometimes end the tour by going for drinks in a quintessentially 'Dalston-style' bar, which has a chequered lino floor (from its previous life as a furniture shop), a glitter ball and lampshades made from plastic colanders.

Dalston is a vibrant area that has undergone dramatic change since the East London rail line was extended to the area. The signs along the main thoroughfare, Kingsland Road, reveal layers of history dating back to when it was a London outpost in the 1800s. Signs from the Art Deco cinemas have been preserved alongside the vernacular D.I.Y. signage of the market stalls and the lettering over the shops and restaurants that reflect the different communities that have made the area home over the years. More recent additions are trendy bars and shops, as the area has become popular with artists and students and it is now an established social destination. Venues continue to pop up and vanish, often retaining the signage and the name of their previous incarnations during their fleeting lifespans.

Opposite: Ilfracombe Type Safari by Karina Monger
Left: Dalston Type Safari by Sarah Hyndman

Type Safari challenge

Take up the challenge and create a typographic composition to reflect the area where you live.

1 Think of a phrase, quote, memory or lyric that you associate with the area, ideally between 10 and 25 characters long (including spaces).

2 Write this down on a piece of paper. Take a pen or pencil with you so you can cross out the letters as you find them.

3 Find the individual letters on the signs that make up your phrase, ideally using a camera with a zoom lens so you can isolate each letter.

4 Have fun and get creative; try daytime and night-time versions for comparison.

5 Combine your photos to spell out your phrase.

What's the difference between legibility and readability?

This is legible but not very readable.	This is legible and much more readable.

Legibility: Can you read the words?

Readability: The choice of typeface, the design and layout can create a more (or less) readable experience.

Answers

Page 16

Anna's family car is a Ford.

Page 18

Supermarket sweep: *There are six luxury products to be found in each supermarket.* Typefaces: *Cooper Black, Edwardian Script, Impact, Helvetica, Franklin Gothic Condensed, Candice.*

Lawyers' typefaces: *Trajan, Stencil, Cinema* italic.

Ampersand pub typefaces: *Engravers' Old English, Flemish Script, Cooper Black.* The Ampersand Arms pub name was suggested by Daniel Smithies via Twitter.

References

1 Sarah Hyndman interviewed by John Humphrys and Justin Webb, 2013, Radio 4 *Today,* typetastingnews.com/2013/09/17/radio-4/.
2 'Food Fight' by Steven Heller, 1999, *AIGA* vol. 17.
3 'Lessons to be learnt from the Gap logo debacle' by Tom Geoghegan, 2010, *BBC News Magazine.*
4 'The road sign as design classic' by Caroline McClatchey, 2011, *BBC News Magazine.*
5 'Why the Same Three Typefaces Are Used In Almost Every Airport' by Alissa Walker, 2014, gizmodo.com.
6 'Where in the world am I? The addictive mapping game that is GeoGuessr' by Will Coldwell, 2013, *Independent.*

2

Functional *vs* evocative

His Consolation of Philosophy printed by William Caxton
in 1478, courtesy of St Bride Library and Archives

Blackletter	Humanist	Old Style	Transitional	Modern	Slab Serif
Ae	Ae	Ae	Ae	Ae	**Ae**
Engravers' Old English	Centaur	Caslon	Baskerville	Didot	Egyptian 710
1450s	1470s	1500s	1700s	1780s	**1800s**

FUNCTIONAL VS EVOCATIVE

Humans are social creatures and the act of communicating with each other, of exchanging experiences and passing on history, has played an important role in our evolutionary survival. The information we share has taught us what is safe and what might be dangerous, helped us to find food and stay healthy, and bonds us together in communities. For millennia this was done through the rich oral tradition of storytelling.

Early books were only available to be read by a privileged few because they were time consuming to create. In the Middle Ages they were meticulously hand-copied by monks, with beautifully ornate detail. These large, heavy tomes were precious and housed in monasteries and universities. The average person had no access to books so reading was not a skill many needed to learn. Instead, the high street signs of the day featured iconic symbols, some of which we still see in use today, like the striped red and white barber's pole, the pawn broker's three hanging spheres, and the visual interpretations of pub names.

Then everything changed...

The introduction of moveable type to Europe meant that multiple copies of books could be printed. This was to be a game changer as books became accessible to the whole population and not just a minority. The first major book to be printed was the *Gutenberg Bible* (see page 126), produced by Johannes Gutenberg in Germany in the 1450s. By the 1470s William Caxton had introduced the printing press to England and he also became the first English book retailer.[1]

Books were still large and heavy, but within a few decades Aldus Manutius[2] invented italic type, which was narrower enabling more words to be printed on a page, and he scaled the size of a book down to make it portable and cheaper to produce. Following these developments a much wider audience gained access to knowledge. Literacy rates increased and education was no longer a luxury available only to the wealthy. Ideas could be shared on an unprecedented scale and a knowledge revolution had begun.

Evolution of type

The first printed typefaces were Blackletter type styles that reflected the meticulous handwriting of the scribes. However, these letterforms were intricate for metal type and the coarse, absorbent paper of the era was not suited to printing such fine details. Typeface design has always gone hand in hand with developments in printing technology and paper quality. In the 1470s printers turned to simpler type styles that were better suited to the process and materials available to them. They took

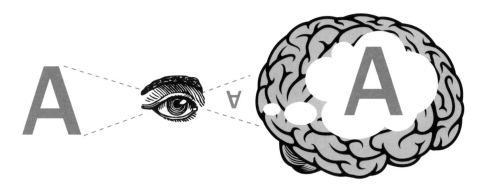
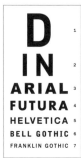

inspiration from the calligraphy of the Italian Humanist writers, who created letterforms closer to the Roman inscriptions found on Trajan's column than the heavy, gothic Blackletter scripts. This classification is known as Humanist[3] and the calligraphic influence can be seen in the shape of the letter 'e' (see opposite).

The transformation from Blackletter script to Roman Serifs was a dramatic change of style. But in the five and a half centuries since there has been relatively little change by comparison, even with the arrival of Sans Serifs. By contrast, fashion and transport have changed beyond recognition from the medieval outfits and horse-drawn carriages of the 1400s to what we wear and the transport options of today.

There is an interesting juxtaposition that takes place when type styles from the 15th and 16th centuries are used on the screens of today's devices. One of the options on a Kindle is *Palatino*[4] which, although designed in the 1940s, is based on the Humanist types of the Italian Renaissance and named after 16th century Italian calligrapher Giambattista Palatino. Yet it does not seem out of place used for 21st century technology. Early printed book specialist Tom Nealon talks of the endurability of Nicolas Jenson's Roman type of the 1470s and says this design had 'nailed it'[5]. He suggests that if you set a book today in Jenson 'no one would bat an eye', yet setting it in Blackletter would be like using 'runes or Klingon'. In the future,

as technology accelerates the pace of change, will typefaces transform beyond recognition or will they continue to reflect these links to history and the printing techniques of the past?

Type is functional

Type functions as a carrier of words. It displays these efficiently so that the reader's eyes can glide seemingly effortlessly across the page as they read. It is sometimes considered that type should be 'invisible' and not intrude on the reading experience. The title of American typographic expert Beatrice Warde's 1930 essay, 'The Crystal Goblet'[6] refers to her opinion that type should function like a clear wine glass and purely 'carry but not obstruct' the content.

Much research into typefaces explores their legibility, focusing on the mechanics of letter shapes and how they function. Testing includes eye-tracking and monitoring response times. An example is the research Monotype type foundry has done with MIT into legibility of typefaces on car dashboards (see page 125).[7]

There are rules for typography, which can be found in many books and blogs on the subject. Over the page are some of the most important ones. (See page 134 for examples and explanations of the terminology).

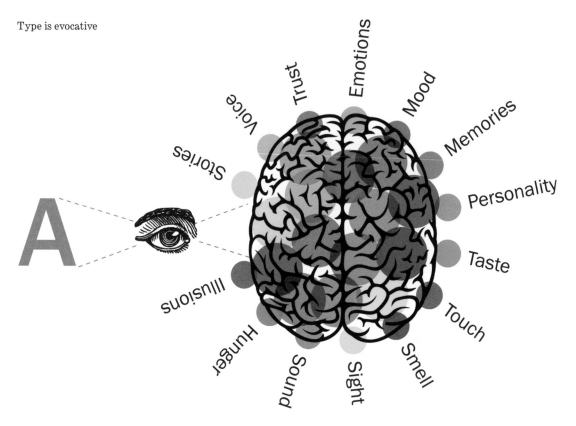

1. The ideal point size for body copy is between 8.5–12 points (print) or 15–25 pixels (web).
2. Line length should be a maxiumum of around 80 characters (including spaces); if your lines are longer then consider more columns.
3. Buy good typefaces for professional work. Don't use free ones as you get what you pay for.
4. Body copy can be difficult to read when centred or justified.
5. Use all caps sparingly: it is less readable and looks like you're SHOUTING.
6. Combining multiple display typefaces can look like you're throwing a fancy dress party.
7. Pale text reversed out of a dark background may fill in when printed and can be hard to read.
8. Never artificially stretch or compress type; choose an extended or condensed font.
9. Beware of bad kerning (the spacing between letters). This can look unprofessional and has been known to create unexpected swear words.

However this book is not about the function of fonts or the rules of typography...

Type is evocative

There is more to type than just being an invisible transmitter of words. The different shapes and styles of the typefaces themselves stimulate responses independently of the words they spell out, and before we even read them.

Type triggers our imaginations, evokes our emotions, prompts memories and links to all of our senses. We automatically recognise attributes from the physical world, like how loud it looks, whether it is heavy or light, fast or slow, or what it would feel like to touch. We have also learned a great deal from our shopping experiences, which include knowing whether something is expensive, aimed at children, or how it might taste.

It is easy to measure how readable a font is, but it is less straightforward to measure how it makes people feel. As a result, there is less public domain research in this area and much of what we know as designers comes from personal observation and experience.

References
1 'William Caxton', en.wikipedia.org.
2 'Aldus Manutius', typographia.org.
3 'History of typography: Humanist' by John Boardley, 2007, ilovetypography.com.
4 'Kindle Paperwhite' by Stephen Coles, 2012, fontsinuse.com.
5 'Kern Your Enthusiasm 12' by Tom Nealon, 2014, hilobrow.com.
6 'The Crystal Goblet' by Beatrice Warde, 1930, World Publishers. 'The Crystal Goblet' is an essay on typography by Beatrice Warde. The essay was first delivered as a speech, called 'Printing Should Be Invisible', given to the British Typographers' Guild at the St Bride Institute in London on October 7 1930.
7 'Monotype Introduces the Burlingame Typeface Family', press release by Monotype 2014, monotype.com.

Online Type Tasting surveys can be found at typetasting.com.

How a typeface functions is practical. It could be compared to building and fine-tuning a car engine for optimum performance. What a typeface evokes is about the experience of being a type consumer and using it. This is like taking the car out for a drive on the open road, turning up the music and enjoying how it makes you feel.

'Typeface' or 'font'?

The TYPEFACE is the design, an example is *Helvetica*, this includes the entire family of sizes, styles and weights.

A FONT is the format you experience the typeface in. Traditionally this would have been a set of metal type in a specific size, style and weight: for example 12-point *Helvetica* bold. Today we access typefaces through many more formats and the meaning of the word 'font' has evolved to encompass them. The term now includes the files we access via the font menus on our computers, mobile devices and websites.

I recently explained this to a client who concluded that 'a typeface is like my collective family, the Greenes, and the fonts would be the individual Greene family members.'

Another comparison could be that a typeface is the story, for example *Star Wars*, and the fonts are the different formats you could choose to watch it in.

The two terms are becoming increasingly interchangeable and I think conversations about type should not be overshadowed by terminology anxiety.

Typeface	*Helvetica*	Greene family	*Star Wars*
Fonts	Helvetica thin, 8 pt *Helvetica thin italic, 8 pt* Helvetica light cond., 8 pt *Helvetica light cond. italic, 8 pt* Helvetica, 8 pt *Helvetica italic, 8 pt* **Helvetica heavy, 8 pt** ***Helvetica heavy italic, 8 pt*** **Helvetica heavy ext.,** ***Helvetica heavy ext.***		

3

How do fonts influence you?

The audience wearing Font Goggles that revealed the 'secret' messages communicated by fonts, at Sarah Hyndman's talk for the London Design Festival at the V&A, 2014. Photo by Martin Naidu

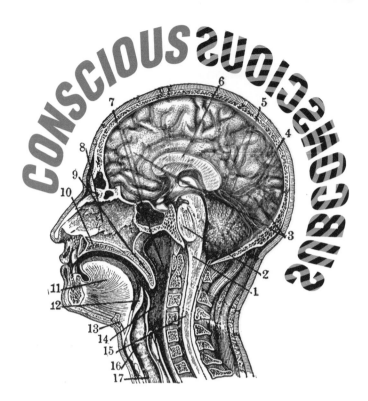

HOW DO FONTS INFLUENCE YOU?

Spoiler alert!
Before reading this chapter take the challenge in this short film: typetasting.com/movie.html[1]

Is type a conspiracy? Does it coerce you with secret messages hidden from view? No, far from it. Type appears right in front of your eyes and is clear for you to see. It is your choice not to pay attention to it consciously and to focus on reading what the words actually say. The type itself is still transmitting plenty of information, but it is communicating it directly to your subconscious.

Through the looking-glass

Typefaces/fonts prime you with a great deal of information and set the scene for the words you are about to read. They give words a backstory, a personality, clue you in to whether or not you can trust them, whether they will be serious or light-hearted, academic or childish. Generally, well-set type is designed so that you 'look past' the typeface and focus on the words themselves — unless of course the font does not match, like somebody with a bad haircut or a miscast actor in a film (see page 56).

At any time you can choose to become consciously aware of the messages the fonts themselves are communicating. On the facing page the red words simply give the name of each typeface, whilst the blue words in 'mirror writing'[2] suggest what they could be saying directly to your subconscious brain. Hold the book up to a mirror to reveal what the blue words are spelling out.

THE FONT
WHAT IS IT TELLING YOU?

Gill Sans

BBC accent

Edwardian Script

Pretentious

Helvetica

Boring*

*allegedly

FRIZ QUADRATA

FRIED CHICKEN SHOP

Comic Sans

!@#?|*c!

Sleight of hand

Neuroscientist Dr David Lewis describes type as being 'hidden in plain sight'[3] and explains that it transmits information 'supraliminally', which means it is within the threshold of awareness. This is the opposite of subliminal, which takes place outside your awareness; for example, it might happen too quickly for you to register it. You can choose whether to notice supraliminal messages, but you are not aware of subliminal ones.

al	Supraliminal	Supraliminal	Supraliminal	Su
E	*BUY ME*	*BUY ME*	*BUY ME*	B
⏱ 0:020	⏱ 0:030	⏱ 0:040	⏱ (
l	Subliminal	Subliminal	Subliminal	S
		BUY ME		
⏱ 0:020	⏱ 0:030	⏱ 0:040	⏱ (

Examples of subliminal messages are the 'blipverts' that featured in the *Max Headroom* sci-fi film in the 1980s. These were television adverts shown so quickly that they influenced viewers subliminally without them realising. At the time these prompted concern that advertisers were really using them to brainwash customers.

Examples of supraliminal communicators are the score or soundtrack to a film, the smell of food, and typefaces, all of which we will look at in more detail in this chapter.

In his article 'The Art of Honest Deception',[4] author Vincent H. Gaddis explains that a magician's skill is not that he can perform tricks too quickly for the eye to see, his skill is in cleverly misdirecting the attention of the audience during the performance. He says that magic tricks are successful because they are 'nine-tenths simple distraction'. Printed words are like magicians, they distract you from paying attention to the fonts and instead you concentrate on reading what they say.

Typefaces and magic tricks both take place right in front of our eyes, but somehow they can seem invisible. Lewis describes this as 'perceptual blindness', explaining that we 'don't perceive what we don't attend to'.[5] Professor Daniel J. Simons

dur du dur du dur du

calls it 'inattentional blindness' for the same reason. Simons demonstrates this well in 'The Monkey Business Illusion'. This is the challenge from the short film at the beginning of this chapter.[6]

Wood for the trees

One of the things I have learnt from my experiences as a graphic designer is to proofread large type in titles and headlines carefully, because this is where spelling mistakes can go unnoticed. It is easy to get so focused on the small details of a design project that we can forget to take a step back and only spot the spelling mistake (with that heart-sinking feeling) once the project is finished and printed. This is a classic example of 'not seeing the wood for the trees'.

I once arranged to meet a friend in Liverpool Street station under the 'really big departures board'. After a while she rang me to say she could not find it because there were too many signs. When I found her I said 'look up' because she was standing right underneath it, but she had

been so busy looking at all the small signs she had not noticed the huge one that ran the full width of the station.

Under the influence

Type sets the tone and gives you clues about what you are about to read. Headlines and titles often feature display typefaces that have larger-than-life characteristics. This works much like the music in a film, which sets the scene for what you are about to watch. The soundtrack tells you whether you can relax, whether to feel sad, or if you should start to feel anxious and get ready to close your eyes before the action actually happens.

It is the iconic music in *Jaws* that creates the tension and primes you to be teetering right on the edge of your seat and ready to jump at the moment the shark actually appears.

In *Alien* the noise gets increasingly frenetic, building up to the shock of the creature suddenly bursting out of John Hurt's chest.

Helvetica **Cheap** *Flemish Script*

Cooper Black
easyJet
VAG Rounded
Volkswagen AG

In the film *The Ring* it is the unanswered phone that makes the scene so creepy when the figure climbs out of the well and then out of the television screen itself. I recently watched this with the sound turned off and not only was it no longer scary, but it became almost comical.

Shops and supermarkets are aware that many factors like type, design, music and aroma can influence how we shop, but without us being aware of this consciously.[7,8] In an article on subliminal selling, Dr David Lewis refers to a study that found that if classic rock was playing in a shop, 'baby boomers' were more likely to make purchases. Yet, when they were asked about it afterwards, two-thirds did not even remember what music had been playing.

In another study C. S. Gulas and C. D. Schewe[9] show that wine buyers will purchase more expensive wine when classical music is playing in a shop. Napa-based graphic designer David Schuemann specialises in designing wine labels and says 'we always make a wine look about $10 more expensive than it is'.[10]

The smell of freshly baked bread or roast chicken wafting through a food shop can have you heading for the bakery aisle or picking up the ingredients for a roast dinner, even if you had no desire to eat either before walking into the shop. I find myself detouring to buy a coffee when I smell fresh coffee or see the font used by my favourite coffee shop.

Arial black

Wine

Didot

Flemish Script

What am I telling you?

Agincourt

What am I telling you?

Didot

What am I telling you?

Helvetica

What are these fonts telling you? Write your answers under each typeface.

Next time you are out shopping and you find yourself with items in your basket that you had not planned to purchase, stop and become aware of all the stimuli around you. Are there external factors influencing your purchasing choices?

Sponsored fonts

When I scroll down my font menu I see some typefaces that I associate with brands. Opinion is divided on whether a company's typeface being in the public domain for everybody to use dilutes its brand image, or whether its increased presence is a clever form of advertising for a strong and recognisable brand.

When I see *Cooper Black*[11] (shown on the previous page) I think of the easyJet logo. The typeface looks to me like comfortable pillow-like clouds and I find I can easily start daydreaming about holidays.

Also in the list is *VAG Rounded*,[12] which was designed for Volkswagen AG in 1979. Making it available to the company worldwide was a problem at the time. To get around this the font was put into

the public domain and was soon included in most free font packages. Although it is no longer their corporate typeface, the link will always be preserved in the name.

I think there is an opening for a big brand to commission a handwritten-style typeface family as an alternative to *Comic Sans* that could be bundled for free with standard operating system software. Since a large proportion of the audience would be schools and children, I would hope it is a brand promoting positive and healthy values, or that encourages curiosity, and not one advertising fast food.

Sinaloa

Wainwright

Madame

References

1 'Monkeying around with the gorillas in our midst: Familiarity with an inattentional-blindness task does not improve the detection of unexpected events' by Professor Daniel J. Simons, 2010, i-Perception. www.youtube.com/watch?v=IGQmdoK_ZfY.

2 'Through the Looking-Glass' by Lewis Carroll, 1871. In the reflected version of her own house Alice finds a book with the poem 'Jabberwocky' written in 'mirror writing', which she could only read by holding it up to a mirror.

3 *The Brain Sell: When Science Meets Shopping* by Dr David Lewis, 2013, Nicholas Brealey Publishing.

4 'The Art of Honest Deception', by Vincent H. Gaddis, 2005, strangemag.com.

5 'Subliminal Selling' by Dr David Lewis, 2013, themindlab.co.uk.

6 Ibid. Professor Daniel J. Simons.

7 Ibid. Dr David Lewis.

8 'Using background music to affect the behaviour of supermarket shoppers' by Milliman, R. E. 1982, *Journal of Marketing*, 46 (3), 86-91.).

9 'Atmospheric Segmentation: Managing Store Image With Background Music' by Gulas, C. S. & Schewe, C.D. (1994).

10 'Drinking With Your Eyes: How Wine Labels Trick Us Into Buying' by Michaeleen Doucleff, 2013, kqed.org.

11 'Cooper Black' by Oswald Bruce Cooper in 1921.

12 'VAG Rounded' by Gerry Barney et al., wikipedia.org.

Online Type Tasting surveys can be found at typetasting.com.

hello **hello** hello

hello **hello** **hello**

hello hello **hello**

hello hello **hello**

hello *hello* hello

hello **hello** hello

hello *hello* hello

hello hello hello

4

Type Karaoke

Part of *Helvetica*'s extensive vocal range

Arkeo

Black Boton

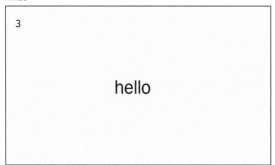

Franklin Gothic

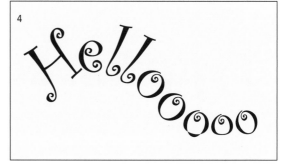

Curlz

TYPE KARAOKE

Play the game of Type Karaoke above by simply saying what you see. Try it out on friends, family or, for an energetic response, with a group of children.

Type can function as a human-voice transmitter and convey the tones and qualities of your voice visually. We each have a unique voice that communicates a great deal about us, like our gender and age. It charts where we come from, who we know and where we have lived along the way. Could your voice look like the traditionally British and approachable *Gill Sans*, or the more intellectual *Caslon*?

Graphic designer Ellen Lupton describes typography as 'what language looks like'.[1] If you imagine that your voice is the typeface, then your vocal range is created by the sizes, styles and weights of fonts within that typeface family and how they are arranged on the page. Even a simple text message can be written expressively, from speaking very s l o w l y to SHOUTING. Think how you feel when you receive a text or email written in all upper-case letters?

Clive Lewis and Peter Walker see the printed word as a visual code for speech that creates a permanent record of both (a) how the words sound when spoken, and (b) what they mean.[2]

In her famous 1930 essay,[3] Beatrice Warde compares typeface legibility to the human voice and suggests that if three pages were set in *Fournier*, *Caslon* and *Plantin* typefaces it would be like 'three different people delivering the same discourse — each with impeccable pronunciation and clarity, yet each through the medium of a different personality.'

Visual onomatopoeia

We all recognise instinctively that type mirrors qualities from the physical world and understanding is consistently similar from one person to the next. A bold font with more ink coverage on the paper looks loud, and if it is large enough to fill the field of vision then it appears so close it comes across as extremely loud. A small word looks a long way away and sounds distant and quiet. The arrangements of

Helvetica Neue ultra light compressed

Flemish Script

Riptide

Modified *Impact*

the letters on the page are easily read, like musical inflection, and you will match your tone of voice to this as you read.

Balega

Type designer Jürgen Weltin says of his 2003 typeface *Balega*[4] that with its contrasting curves and sharp edges it would ring with the 'fat sound of a Jimi Hendrix guitar.'

On page 41 there are some of the descriptions from a Type Tasting online survey which asked what typefaces would sound like if they were music?

Gill Sans

I always imagine that *Gill Sans*, designed in the 1920s, would speak with a 'BBC English' accent. It would pronounce words and use grammar correctly, but have a friendly and relaxed tone, a few notches down from the formality of the Queen's English. It is the typeface used for the BBC logo and is reminiscent of 1930s 'keep calm' England.

Type and Sound workshop

So far in this chapter we have explored what existing fonts would sound like. In one Type Tasting 'Type and Sound' workshop, participants take sound as the starting point and go through the process of using it to inspire new letterforms. The final alphabets demonstrate how letterforms can convey the nuances of different sounds.

Participants are played three different sounds in turn. These are heavy metal music, waves on a beach and a screaming woman from a horror film. As they listen they fill a sheet of paper with scribbles and marks inspired by each sound using appropriate materials.

Once the three sheets have been filled with marks the participants search for letterforms in each and cut them out. At the end they have created three distinct alphabets, each of which reflects the qualities of the sounds that inspired it.

See Lydia CS's interpretations over the page.

Type and Sound workshop
Sketches and sound alphabets (A to E) by Lydia CS

Heavy metal music

Waves on a beach

Screaming woman

Type Tasting online survey in which participants are asked what typefaces would sound like if they were music?

sound

Robot, retro, techno, crisp, digital, discordant, edgy, futuristic, clipped, green, bright, icy, juttery, solid, snappy, firm, trebly, edgy, resonant, tinny, atonal, loud, cacophonous, cutting, clipped, edgy, deep, robot, classic, booming, quick, repetitive, auto-tuned, soulful, techno, technical, modern, digitised, retro, underground, boxy, sci-fi, electronic, techno, ugly, thematic, technical, beatbox, electric, '80s, funky, different, space-age, bleak, haunting, quirky, artificial, techno-beat, industrial, serious, oppressive.
It makes me feel anxious.

Cinema

sound

Sophisticated, mellifluous, melodious, relaxing, classic, mellow, smooth, deep, mature, warm, rounded, round, soft, smooth, heavy, comfy, curved, deep, bass, bassy, melodious, loud, low, blue, melodic, slow, fluid, bouncy, gentle, retro, mature, sure, sophisticated, soul, classic, full, soulful, classy, psychedelic, old, complex, austere, jazzy, traditional, ice cream, dated, modern, summery, sexy, romantic, relaxing, happy, pleasing, mellow, calm, mellow, fun, chilled, serene, immersing.
It makes me feel happy and calm.

Bodoni Poster italic

sound

Jarring, sharp, energetic, harsh, tinny, edgy, dramatic, piercing, uncomfortable, modern, cool, cold, bright, vibrant, scratchy, spiky, prickly, jagged, angular, tinny, high-pitched, jangly, percussive, loud, pitchy, shrill, discordant, piercing, high, stuck, fast, quick, energetic, staccato, brief, erratic, jerky, jittery, strict, punk, funky, modern, quirky, electronic, futuristic, urban, repetitive, rave, ear-piercing, different, noisy, synthesised, dramatic, young, angst, agitating, exciting, uncomfortable, aggressive.
It makes me feel anxious.

Modified *Klute*

sound

Safe, young, playful, soft, boring, balanced, childish, calming, straightforward, sonorous, fresh, cool, vibrant, heavy, rounded, soft, round, wooden, solid, smooth, tympanic, full, resonant, even, smooth, clear, ringing, melodic, loud, booming, ambling, fluid, bouncy, breezy, mundane, simple, standard, perfect, clean, classy, clear, neat, friendly, basic, easy, simple, child-like, clichéd, bold, boring, clear, safe, comforting, relaxed, upbeat, pleasant, happy, confident, sad, calming, playful, gentle.
It makes me feel happy and calm.

VAG Rounded

References
1 'Thinking with Type' by Ellen Lupton, 2010, Princeton Architectural Press.
2 'Typographic influences on reading' by Clive Lewis and Peter Walker, 1989, *British Journal of Psychology*.
3 'The Crystal Goblet' by Beatrice Warde, 1930, World Publishers.
4 'Balega' by Jürgen Weltin, 2003, membership.monotype.com.
Online Type Tasting surveys can be found at typetasting.com.

(a)

(b)

(c)

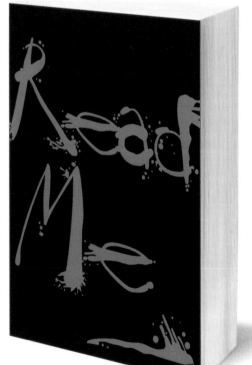

(f)

(g)

(h)

5

Fonts turn words into stories

(d)

(e)

Book genres by type
(a) *Slipstream,* (b) *Wainwright,* (c) *Arnold Boecklin,* (d) *Engravers' Old English,* (e) *Flemish Script,* (f) *Bludgeon,* (g) *Retro* Bold, (h) *Comic Sans,* (i) *Data 70,* (j) *Stencil.*

(i)

(j)

A font can transform the meaning of words

SALESMAN *Wainwright*

Ancient Worlds *Amelia*

Macho Man *Stilla*

Serial Killer *Edwardian Script*

HIGH TECH *Trajan*

Outer Space *Engraver's Old English*

FONTS TURN WORDS INTO STORIES

'Come with me and you'll be in a world of pure imagination.' Willy Wonka.[1]

Typography is storytelling. Typefaces set the scene and clue you in to what the words will reveal, independently of what they actually say. This is something referred to by psychologists Clive Lewis and Peter Walker as 'typographic allusion'.[2] By doing this, fonts give words a backstory and a personality; they establish meaning and sometimes they subvert the meaning of the words. Your interpretation of a word can be brought to life by the typeface in which it is set.

There is a coffee shop I go to that has one wall completely lined with books with only the spines on view. The book jackets have been reduced down to typography and colour, with all the different styles giving tantalising glimpses of the stories to be found inside the books. Often it is the typeface that draws my eye to a particular book and makes it look so much more interesting than the one sitting next to it.

Responses, associations or knowledge?

1 Instinctive responses
You react instinctively to some typefaces and respond purely to their shape. Many responses are hardwired into your DNA for your survival and protection. For example, you intuitively know that round shapes are safe and friendly, whilst jagged shapes are dangerous or aggressive (see page 69).

2 Learned associations
You have a library of associations in your brain that you have been collecting all your life. This happens every time you see a typeface from the context in which you see it used. Some of these can be almost universally recognised; others are specific to you and your own unique experiences.

These associations can remain constant, unaltered over time — or they can change with the flick of an event that adds to their history and creates a completely new set of associations. For example, Blackletter was used for the first books printed in Germany in the 1400s. These styles were based on the handwriting of monks and have associations with timeless wisdom and precious books like The Bible. However, when Hitler declared Blackletter to be the official type style of the Nazi Party, this ancient and romantic type style acquired a new and darker set of associations that have become permanently entwined with its history.

Tobias Frere-Jones designed the *Gotham* typeface family for *GQ* magazine. It is based on the mid 20th century geometric type found on vernacular signage in New York.[3] When *Gotham* was used to spell out 'hope' and 'change' in Obama's 2008 presidential campaign it became the typeface of 'inspiration and optimism' and is now so widely used that it has joined the ranks of the celebrity typefaces.

'No typefaces are immune to history' Neville Brody.[4]

Since we gather these learned associations throughout our lives, older generations could have a much broader frame of reference for type as they have spent 'many more years becoming acclimatised to the links between fonts, products and experiences', according to journalist Alex Finkenrath. Just take a look at other areas of popular culture, like music, where parents of every generation have uttered the phrase, 'I had the original version of that song' to a child who thinks they are listening to something brand new.

3 Learned knowledge
This is the domain of the expert who, like an art critic or a wine connoisseur, has acquired a great deal of knowledge of the subject. The expert generally interacts with type as a professional designer and not just as a consumer, and will refer to a complex set of qualities and associations learned throughout their career (see page 81).

A font is in the eye of the beholder

As graphic designers we take advantage of these links and associations. We use these visual codes to enable us to communicate messages effectively and to create layers of meaning. By doing this we also reinforce and strengthen this code of references through repetition.

Designers combine typefaces with words using different levels of complexity.

1 The font matches the words
The advantage of using a type style that mirrors what the words say is that the words can be understood easily from a quick glance. There may be occasions where this is important.

Wainwright

Edwardian Script

2 The font adds extra meaning to the words
The typeface works a little harder and conveys

extra information that adds extra meaning to the words. The western below is now set in Mexico and the romance has become an army adventure story.

Mesquite

Stencil

3 The font alters the meaning of the words
Now the western is set in the sci-fi world of outer space and the romance is clearly not destined for a happy ending.

W E S T E R N

Futura

Romance

Bludgeon

Moving away from clichés

The 'Read Me' books on pages 42–43 have covers that epitomise particular genres. Which cover fits each of these categories: 1960s sci-fi, vampire story, comedy, action thriller, romance in Paris, army adventure, classic romance, rebel against society, western, horror (answers on page 53).

Moving away from clichés enables the type to create intrigue and to become more complex by adding layers of meaning. The typeface can allude to additional ideas and the designer now relies on the reader's knowledge of these references to recognise and understand the combined meanings. The word treatments at the top of page 44 show how fonts can transform the meanings of words to suggest more complex stories.

Do you believe the type?

As designers we can use type to imply things that we would not be able to say in words, or which might be untrue if shown as a photograph. For example, if a beef burger is mass-produced in a factory it would be a lie to say 'hand-made' on the packaging. It would also be untrue to show an image of somebody making it by hand. However, we can use a typeface that implies the product is hand-made and the consumer may assume it is an

(a)

Comedy

(b)

Romantic comedy

(c)

ANY MOVIE

accurate description of the product. This is a widely used advertising technique. Do you think you have ever fallen for it?

BURGER

BURGER

Compare the two typefaces above: which product do you assume is hand-made? Which would you pay more for? You may even find yourself enjoying the taste of the second burger more because you expect to, even if they are both the same burger (see page 115 for more about primed experiences).

Semiotics and film posters

There are continuing discussions about whether type should simply be an unseen transmitter of words (it was mentioned earlier (page 25) that Beatrice Warde taught that type should be 'invisible'.[5]) I think well-designed type can appear transparent and not intrude on the reading experience, but that it does not remain neutral.

Semiotics, which became popular in the 1960s, is the study of signs and meanings.[6] According to its principles, an object becomes inextricably linked with ideas or associations you make from the context in which you encounter it. Not only does the object lose its neutrality, it also goes on to become a transmitter of the additional meanings. Both the object and the meaning combine to become a 'sign'. For example, a red rose is linked to the idea of romance and, once you have made this connection, it is hard to think of one without the other.

Writer Yves Peters researches the typography of film posters and explains that some typefaces have come to signify particular film genres because they have been used so repetitively that they have created a graphic code (see above).[7] It is difficult to look at these typefaces without the associated genre coming to mind.

(a) Comedy film titles are often in bold Sans Serif typefaces like *Gill Sans* ultra bold.

(b) Romantic comedies generally use Modern/Didone typefaces such as *Didot*.

(c) The typeface *Trajan* was first used for epic inspirational films, but it has now appeared so frequently across all genres that it has become the ubiquitous typeface of the movies. It was even adopted by the Oscars in 2006.

In The
Blackmoor

Beginning
Goudy Text

Type
Agincourt

emulated the
Luthersche Fraktur

handwriting
Engravers' Old English

of scribes
Fette Gotisch

Humanist

E.g. *Centaur*

circa 1470 · de — Calligraphic influence / sloping crossbar on 'e' / low thick/thin contrast / sloped stress

Italic

E.g. *Garamond italic*

circa 1500 · de

Old Style

E.g. *Caslon*

circa 1500 · de — Horizontal crossbar on 'e' / greater thick/thin contrast / more upright stress

Transitional

E.g. *Baskerville*

circa 1700 · de — More refined serifs / greater thick/thin contrast / upright stress

Modern

E.g. *Didot*

circa 1780 · de — Abrupt (unbracketed) serifs / extreme thick/thin contrast / upright stress

Slab

E.g. *Egyptian 710*

circa 1800 · de — Bold 'slab'-shaped serifs

In the beginning

The first printed typefaces were created in the 1450s. These were based on the detailed writing of the monks and scribes who hand-copied books before the advent of print. Our modern eye can find these styles overly ornate and difficult to read in quantity, but if you agree with type designer Zuzanna Licko's premise that you 'read best what you read the most'[8] then Blackletter could be thought of as the *Helvetica* of its day.

Blackletter has absorbed many associations in its long history from the different contexts in which you have seen it used. These include the romance of Arthurian legend, the rebellion of heavy metal music, the links to tradition through newspaper mastheads, the darkness of vampire films and the authenticity of German beer labels.

Traditional and knowledgeable

Serif typefaces date back over five centuries to the first decades of printing in Europe and they have become steeped in associations of history, knowledge and tradition. Many books, especially older books, are set in Roman Serif styles, as are traditional newspapers. As a result we have come to associate these with conveying serious and credible information.

Anna Alternates

Anna Extended

CELEBRATED

Huxley Vertical

LUXURY

Sinaloa

GLAMOUR

Copasetic

AND THE

Matra

MACHINE AGE

Plaza Swash

Nostalgic luxury

Art Deco became popular in the 1920s and 1930s. It stood for modernity and progress and offered an escape from traditional conventions.[9] Many cinemas and music halls were built in England in this era and both the architecture and signage can still be seen. These type styles are popular on bar signs for their associations with the decadence of speakeasies in the prohibition era.

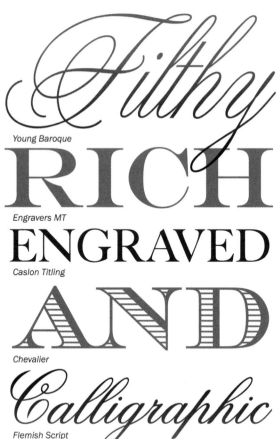

Young Baroque

RICH

Engravers MT

ENGRAVED

Caslon Titling

AND

Chevalier

Calligraphic

Flemish Script

Money, money, money

The typefaces above are all about money and are inspired by the style of engraved typography found on bank notes, deeds and bearer bonds. Formal scripts similar to *Young Baroque* are based on the skilled letterforms of 18th-century master calligraphers, who used quills or metal-nibbed pens. The Roman-style letters, like *Chevalier*, have fine lines, details and sharp serifs typical of engraved styles.

The intricate designs on bank notes are engraved by hand onto metal plates by engravers like Tony Maidment. This is a highly skilled craft and the work by each engraver is individual, like handwriting or a fingerprint. The nuances of lines and dots, unique to the style of each engraver, are one of the main weapons against counterfeiting.[10]

F

Didot Headline

IS FOR

Bodoni

fashion

Didot

type for a

Industria

REVO-LUTION

Retro Bold

BREAKS FREE FROM

Pritchard

THE GRID

East Bloc Phosphor

Fashion fonts

Typefaces with extremely contrasting thick and thin elements are classified as Moderns, or Didones. These were developed in the late 1700s by famous rivals Firmin Didot in France and Giambattista Bodoni in Italy. Improvements in paper quality and printing technology enabled them to print letters with extremely fine details and crisp, sharp edges.

These have been used widely on fashion magazines like *Vogue* and *Harper's Bazaar* since the 1920s and, as a result, we now think of them as epitomising femininity, style and fashion.[11, 12]

Revolutionary type

Avant-garde art movements of the late 19th and early 20th centuries like Constructivism, De Stijl, Futurism and Dadaism, have dynamic visual styles that treat type as imagery, allowing it to break free of the structured grid.

These were the movements of revolution and change; they pushed boundaries and embraced dynamic progress. They celebrated noise, momentum and energy.

Famous avant-garde designers who worked with type include El Lissitzky, Alexander Rodchenko and Filippo Tommaso Marinetti.

RULE

Gill Sans Shadow

Brittania

Johnston

from Baskerville

Baskerville

& Caslon to the

Caslon italic '&', Caslon

GREAT

Festival Titling

humanist

Gill Sans heavy

sans serifs

Johnston bold

Made in Britain

This collection of typefaces is a celebration of Britishness. *Gill Sans* is a Humanist Sans Serif with a hand-drawn feel to its shapes. It was designed by the artist and type designer Eric Gill. Gill was a student of Edward Johnston, creator of *Johnston* for the London Underground. (In my opinion *Johnston* is the unofficial typeface of London.) *Baskerville* and *Caslon* represent the rich British history of type design, which dates back to the 1600s, with typefaces such as these exported around the world. *Caslon* was used for the Declaration of Independence. *Festival Titling* was the typeface created for the 1951 Festival of Britain and it still evokes 1950s nostalgia.

Appleton

Victorian

Algerian

Friz Quadrata

AdHoc

Brody

Flash

Writing on the wall

Painted lettering on signs became popular in England in the 18th century with the increase in literacy following printed books and newspapers becoming more widely available.[13] Signwriters were the original 'brand designers', as they created the signage that advertised shops and businesses. There is a current resurgence of interest in the craft of signwriting, with many more signs being traditionally hand-painted. You can often recognise which craftsman's work it is from signature details like a flourish on the crossbar of the 'A'.

A

Blackoak

New era of

Egyptian Slate

ADVERTISING

Wainwright

&type

Falstaff

WENT

Enge Holzschrift

LARGE

Magnifico Daytime

THE

Broadcast Titling

VICTORIANS

Aesthetique

HAD A

Thunderbird

TASTE

Madame

FOR THE

Wood Relief

Highly

Victorian

ORNATE

Saraband Initials

Going large

The Industrial Revolution in the late 18th century enabled products to be manufactured in excess of demand. The advertising industry was born as manufacturers now needed to promote and sell products. Posters and advertisements had to stand out in crowded environments, on the sides of moving trains, or viewed at angles, and so bigger and bolder display typefaces were created.

Fat Face typefaces, such as *Falstaff*, are bloated versions of Modern/Didone styles (see page 49). Serifs were then enlarged and made chunkier to create Slab Serif typefaces such as *Egyptian Slate*. Later, condensed Sans Serifs and 3D layered type were added into the mix.

Ornamental type

In the 19th century the Victorians embraced ornamentation and lavishly decorated much of what surrounded them. This included fashion, furniture, architecture and also typeface design. Signwriters added flourishes and textures like gold leaf, and printers carved increasingly ornate letters creating three-dimensional illusions and often featuring motifs of everyday life and nature. Some of the most highly ornamented were created by Louis John Pouchée[14] (see page 76) and can be viewed in the St Bride Library in London .

OUT IN
Eurostile Extended 2

SPACE
Sonic XBd

TYPE IS
Futura light

GEOMETRIC
Dex Gothic

sometimes
retro
Amelia

AND APPEARS
Space

BUILT BY
Flatiron

MACHINE
Computer

Modernism
Helvetica Neue medium

Change
Univers black

&
Neue Haas Grotesk black

Neutrality
Univers light

Function
Neue Haas Grotesk light

NOT
Neue Haas Grotesk bold

Ornamentation
Helvetica Neue italic

Intergalactic communication

Many sci-fi typefaces are based on geometric shapes. These look mechanically constructed, giving the impression they were created in a future 'machine age'. They incorporate a graphic language that can imply an underlying theme: the type can suggest an optimistic future utopia, a dark dystopia, or a retro version of the future. The typeface *Amelia* looks like it's been beamed directly from the 1970s. *Eurostile* and the geometric *Futura* have both appeared in so many sci-fi films that they have become the 'corporate fonts' of the genre.

Style follows function

In the first half of the 20th century, graphic design combined Serif typefaces with images and ornamentation that often filled all of the space on the page. With the arrival of Modernism in the 1950s design was simplified; it employed Sans Serif typefaces arranged to a grid structure, with lots of white space. Unnecessary details were removed and style was led by function. Sans Serif typefaces were chosen for their neutrality, because they were new and not steeped in the associations of history. The ubiquitous *Helvetica* (originally *Neue Haas Grotesk*) was designed during this era and has since reached celebrity status, even starring in its own film.

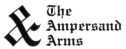

Answers: do you agree?
(a) Action thriller, (b) western, (c) romance in Paris, (d) vampire story, (e) classic romance, (f) horror, (g) rebel against society, (h) comedy, (i) 1960s sci-fi, (j) army adventure.

References
1 *Willy Wonka and the Chocolate Factory*, by Leslie Bricusse & Anthony Newley, 1971.
2 'Typographic influences on reading study' by Clive Lewis and Peter Walker, 1989, *British Journal of Psychology*.
3 'The Origin of Gotham' by Hoefler & Co., typography.com.
4 'Unquiet Film Series: Times New Roman' by *The Times* and *The Sunday Times*, foreverunquiet.co.uk.
5 'The Crystal Goblet' by Beatrice Warde, 1930, World Publishers.
6 'Semiotics for Beginners' by Daniel Chandler, visual-memory.co.uk.
7 'Trajan in movie posters: the rise and fall of the Roman Empire' by Yves Peters, 2011, Beyond Tellerrand Conference.
8 'Interview with Zuzana Licko' by Rudy VanderLans, 1984, *Emigre*.
9 'Art Deco Around the World', V&A.
10 'Feature: The secret art of the engraver' by Mark Sinclair, 2011, *Creative Review*.
11 'Through thick and thin: fashion and type' by Abbott Miller, 2007, *Eye Magazine*.
12 'A Brief History of Type Part Four: Modern (Didone)' by John Boardley, ilovetypography.com.
13 'The English Signwriting Tradition' by Richard Gregory, signpainting.co.uk.
14 'Pouchée's lost alphabets' by Mike Daines, 1994, *Eye*.

Online Type Tasting surveys can be found at typetasting.com.

1

PATRICK BATEMAN

212.555.6342
PIERCE & PIERCE

WHAT JOB DO I DO?

- -

2

Patrick Bateman

212.555.6342
Pierce & Pierce

What job do I do?

- -

3

Patrick Bateman

212.555.6342
Pierce & Pierce

What job do I do?

- -

4

Patrick Bateman

212.555.6342
Pierce & Pierce

What job do I do?

- -

5

Patrick Bateman

212.555.6342
Pierce & Pierce

What job do I do?

- -

6

Patrick Bateman

212.555.6342
Pierce & Pierce

What job do I do?

- -

7

Patrick Bateman

212.555.6342
Pierce & Pierce

What job do I do?

- -

8

Patrick Bateman

212.555.6342
Pierce & Pierce

What job do I do?

- -

6

Don't believe the type

Play the 'What job do I do?' business card game on the left.
Turn to page 84 to compare your answers with the Type Tasting survey responses.

1 *Trajan*, 2 *Monotype Corsiva*, 3 *Cocon*, 4 *Clarendon*, 5 *Bodoni Poster*, 6 *Didot*, 7 *Comic Sans*, 8 *Cinema* italic

rat
tortoise
heavy
light
strong
light
heavy

fast
slow
elephant
weak
gazelle
weak
elephant

fast
gazelle
tortoise
slow
strong
rat

Quiz: Speed Test

DON'T BELIEVE THE TYPE

You have an intuitive sense of when type fits the situation and when it doesn't. When the type is appropriate to the content it enhances the reading experience, which becomes seemingly effortless as a result. If you watch a film with well-cast actors you are able to sit back, suspend disbelief and enjoy it. If they are poorly cast then the film will not ring true and this can spoil even a great story. This does not mean that the actors go unseen, it means that they complement the scenario. In the same way, well-designed type is not invisible, it works in harmony with the content.

Processing fluency

When you see something written in an unfamiliar typeface your reading experience is interrupted because you need to pay attention to deciphering the letters. This means that you read more slowly and you are likely to become aware of the reading process. By contrast, when you read a familiar type style your eyes are able to 'skim effortlessly across the text' and, according to Dr David Lewis, this earns your trust.[1] He calls this 'processing fluency'

as it describes the ease with which you recognise and comprehend what the words say. He goes on to explain that the more easily a customer can process and understand a sales message, the more likely they are to purchase an item. Processing fluency applies not only to the readability of the typeface itself, but also to the design, layout and language. If a highly readable typeface is poorly typeset and difficult to read it will still slow the reader down.[2]

However, this is not proposing that 'one font fits all', as different typefaces will suit different scenarios. The ease with which you read a typeface is a combination of it being both familiar and fitting the context.

Quiz: Speed Test

Set a timer or stopwatch for 30 seconds. Start the timer and circle all the words at the top of this page where the font DOES NOT match the word. That is, where the meaning of the word seems opposite to the style of the typeface (for example, if the word 'fast' is in a typeface that looks heavy and slow). How many did you find? The answers are on page 63.

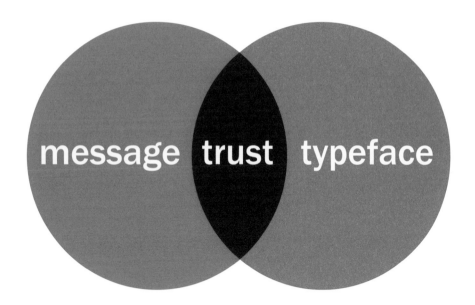

This speed test quiz is inspired by a study called 'Typographic influences on reading', by Clive Lewis and Peter Walker[3], in which participants were shown a selection of words like 'elephant' and 'fast' typeset either in chunky *Cooper Black,* or the light and speedy looking *Palatino* italic. These are the two typefaces used in the Speed Test quiz. The meanings of some words mirrored the shapes of the typefaces and, as in the quiz, some appeared to be the opposite. Participants were given specific words to look out for and asked to press a button as soon as they saw them.

It was found that they were able to identify the words much faster when the words matched the fonts, and that reaction times were considerably slower when they did not match. Lewis and Walker were able to demonstrate that we read much faster when typefaces match the content.

Influential types

In the world of business some typefaces are more appropriate for certain professions than others, and you are more likely to trust a business card in an appropriate style. On page 18 you selected the lawyer you would hire from either (a), (b) or (c) (below). Would you have made a different choice if you had wanted a quick and discount lawyer for something routine and unimportant?

LAWYER **LAWYER** *Lawyer*

(a)   (b)   (c)

For lawyers (d) to (f) below, the choice is less clear cut because all three look appropriate for the legal profession. You now need to make your decision based on personal preference and research.

LAWYER *Lawyer* LAWYER

(d)   (e)   (f)

Many professions have recognisable typeface conventions, especially established industries where the preference is likely to be for traditional, conservative and relatively neutral typefaces that suggest history. Often these have links to Roman inscriptions, engravings or ancient manuscripts that give the impression of longevity and wisdom.

The News

ANCIENT CAT FOUND ON MARS

The News

Ancient Cat Found On Mars

Lorem ipsum dolor sit amet, at nam tempor deterruisset, sonet facete verear eu mei, cum ad dicam ubique discere. Animal definitiones usu cu, dicat probatus at eum. Ex nominavi aliquando pri, tation altera lobortis ut sit. Te mei possit praesent, no timeam commune intellegat his, reque moderatius an mel. At graece animal mea. Sententiae theophrastus usu an, per omnium assentior eu. Mei errem munere vituperata id. Cum ne maiorum alienum, duo tation laboramus in. Soluta prompta vel in, ad his prima error. Id nec minimum omnesque apparaet, te alii tollit decore mei, ignota evertitur rationibus mea ex. Mea habeo bonorum et. Ea quas illud prodesset vix, est ad eruditi fuisset, est at theophrastus definitiones. Habeo iisque veritus mea et. Est cu hendrerit

Mei errem munere vituperata id. Cum ne maiorum alienum, duo tation laboramus in. Soluta prompta vel in, ad his prima error. Id nec minimum omnesque apparaet, te alii tollit decore mei, ignota evertitur rationibus mea ex. Lorem ipsum dolor sit amet, at nam tempor deterruisset, sonet facete verear eu mei, cum ad dicam ubique discere. Animal definitiones usu cu, dicat probatus at eum. Ex nominavi aliquando pri, tation altera lobortis ut sit. Te mei possit praesent, no timeam commune intellegat his, reque moderatius an per omnium assentior eu. Mei errem munere vituperata id. Cum ne maiorum alienum, duo tation laboramus

Mea habeo bonorum vix, est ad eruditi definitiones. Habeo hendrerit interpret mediocritatem in. E eros porro at pri, s pro te. Mea habe prodesset vix, est theophrastus defin usu an, per omnium Mei errem munere mea et. Est cu hend referrentur mediocr iracundia, eros p comprehensam pri Sententiae theoph assentior eu. Mei Cum ne maiorum a

Lorem ipsum dolor sit amet, at nam tempor deterruisset, sonet facete verear eu mei, cum ad dicam ubique discere. Ex nominavi aliquando pri, tation altera lobortis ut sit. Te mei possit praesent, no timeam commune intellegat his, reque moderatius an mel. At graece animal mea. Sententiae theophrastus usu an, per omnium assentior eu. Mei errem munere vituperata id. Cum ne maiorum alienum, duo tation laboramus in. Soluta prompta vel in, ad his prima error. Id nec minimum omnesque apparaet, te alii tollit decore mei, ignota evertitur rationibus mea ex. Mea habeo bonorum et. Ea quas illud prodesset vix, est ad eruditi fuisset, est at theophrastus definitiones. Habeo iisque veritus mea et. Est cu hendrerit interpretaris, eam solum referrentur mediocritatem

Mei errem munere vituperata id. Cum ne maiorum alienum, duo tation laboramus in. Soluta prompta vel in, ad his prima error. Id nec minimum omnesque apparaet, te alii tollit decore mei, ignota evertitur rationibus mea ex. Lorem ipsum dolor sit amet, at nam tempor deterruisset, sonet facete verear eu mei, cum ad dicam ubique discere. Animal definitiones usu cu, dicat probatus at eum. Ex nominavi aliquando pri, tation altera lobortis ut sit. Te mei possit praesent, no timeam commune intellegat his, reque moderatius an mel. Sententiae theophrastus usu an, per omnium assentior eu. Mei errem munere vituperata id. Cum ne maiorum alienum, duo tation laboramus vel in, ad his prima error. Id nec minimum omnesque apparaet, te alii tollit decore mei, ignota evertitur

Mea habeo bonorum et. Ea quas illud prodesset vix, est ad eruditi fuisset, est at theophrastus definitiones. Habeo iisque veritus mea et. Est cu hendrerit interpretaris, eam solum referrentur mediocritatem in. Ea vim erant facilisis iracundia, eros porro at pri, suas copiosae comprehensam pro te. Mei errem munere vituperata id. Cum ne maiorum alienum, duo tation laboramus in. Soluta prompta vel in, ad his prima error. Id nec minimum omnesque apparaet, te alii tollit decore mei, ignota evertitur rationibus mea ex. Lorem ipsum dolor sit amet, at nam tempor deterruisset, sonet facete verear eu mei, cum ad dicam ubique discere. Ex nominavi aliquando pri, tation altera lobortis ut sit. Te mei possit praesent, no timeam commune intellegat his, reque moderatius an

(a)   (b)

However, a good designer can be inventive and use more modern or unexpected typefaces in a way that still feels sympathetic and appropriate to the profession, moving away from stereotypes.

EVERYTHING NOT SAVED WILL BE LOST

NINTENDO 'QUIT SCREEN' MESSAGE [4]

Setting this ephemeral Nintendo game 'quit screen' message in *Trajan,* which is a typeface inspired by Roman inscription, gives it an almost philosophical air of gravitas and longevity.

Which of the two newspaper headlines above are you more likely to believe? Newspaper (a) has been set in the blocky Sans Serif letter style of a tabloid or scandal sheet. These are described by Steven Heller and A. M. Cassandre in *Eye*[5] as the 'screaming headlines' of tabloid newspapers. The assumption would be that the headline is designed to grab attention but the story is likely to be for entertainment value and less likely to be based on fact. The typeface used here is *Franklin Gothic*

bold condensed. Newspaper (b) has the Roman Serifs of a more 'intellectual' newspaper in which you would expect the article to be well-researched and based on fact. This is *Times New Roman*, which was commissioned by *The Times* newspaper in 1931.

Which is the most believable typeface?

In 2012, writer Errol Morris ran an online experiment with the *New York Times*.[6] He wrote an article titled 'Are You an Optimist or a Pessimist?' and at the end he asked readers to score how highly they agreed with it; in other words, how believable they found it.

What readers did not realise was that the article was presented to them in one of six randomly assigned typefaces. Around 45,000 people responded and, after analysing the results, Morris was able to declare *Baskerville* to be 'the most believable typeface'. It was only a 1.5 per cent difference that put it ahead, but he considered this to be a significant difference if you thought of it in terms of votes or sales.

Imperial		New York Times
Baskerville		New York Times
Computer Modern		New York Times
Georgia		New York Times
Trebuchet		New York Times
Helvetica		New York Times
Comic Sans		New York Times

Baskerville
Lorem ipsum dolor sit amet, consectetuer adipiscing elit, sed diam nonummy nibh euismod tincidunt ut laoreet dolore magna aliquam erat volutpat. Ut wisi enim ad minim veniam, quis nostrud exerci tation ullamcorper suscipit lobortis nisl ut aliquip ex ea commodo consequat.

Computer Modern
Lorem ipsum dolor sit amet, consectetuer adipiscing elit, sed diam nonummy nibh euismod tincidunt ut laoreet dolore magna aliquam erat volutpat. Ut wisi enim ad minim veniam, quis nostrud exerci tation ullamcorper suscipit lobortis nisl ut aliquip ex ea commodo consequat.

Georgia
Lorem ipsum dolor sit amet, consectetuer adipiscing elit, sed diam nonummy nibh euismod tincidunt ut laoreet dolore magna aliquam erat volutpat. Ut wisi enim ad minim veniam, quis nostrud exerci tation ullamcorper suscipit lobortis nisl ut aliquip ex ea commodo consequat.

Trebuchet
Lorem ipsum dolor sit amet, consectetuer adipiscing elit, sed diam nonummy nibh euismod tincidunt ut laoreet dolore magna aliquam erat volutpat. Ut wisi enim ad minim veniam, quis nostrud exerci tation ullamcorper suscipit lobortis nisl ut aliquip ex ea commodo consequat.

Helvetica
Lorem ipsum dolor sit amet, consectetuer adipiscing elit, sed diam nonummy nibh euismod tincidunt ut laoreet dolore magna aliquam erat volutpat. Ut wisi enim ad minim veniam, quis nostrud exerci tation ullamcorper suscipit lobortis nisl ut aliquip ex ea commodo consequat.

Comic Sans
Lorem ipsum dolor sit amet, consectetuer adipiscing elit, sed diam nonummy nibh euismod tincidunt ut laoreet dolore magna aliquam erat volutpat. Ut wisi enim ad minim veniam, quis nostrud exerci tation ullamcorper suscipit lobortis nisl ut aliquip ex ea commodo consequat.

However, I think it is important to take context into consideration before crowning *Baskerville* as the king of believability. Trust is created when the typeface matches the content and this suggests to me that the readers who read the article set in *Baskerville* found it to be the most authentic and credible for the *New York Times*. There will be other situations in which it would not score so highly. The newspaper's long-established typeface is *Imperial*, which it has used since 1967.

Can a typeface make you appear more intelligent?

Georgia
23 essay average
A

Times New Roman
11 essay average
A-

Trebuchet
18 essay average
B-

When student Phil Renaud was nearing the end of his third year at university he noticed that his grade average had improved. He wondered why, since he did not think he was putting any more effort into studying or writing. He realised that the one thing he had changed over time was his choice of font, and so he looked back at the 52 essays he had submitted and compared the grades and typefaces. He found that when he used *Georgia* his grade average was A, whilst the essays written in *Trebuchet* only averaged B-minus.[7]

Serif typefaces are associated with academia and knowledge, whereas Sans Serif *Trebuchet* does not convey the same timeless gravitas. Out of the two Serif style typefaces the difference may be due to readability. *Times New Roman* was designed to fit into the narrow column widths of *The Times* newspaper, but reads less comfortably spread across the full width of a page. The more open letters of *Georgia* flow better across a wide page and read more smoothly. I suggest that *Georgia* was favoured because it combines appropriateness for the context with being a good reading experience.

Try this out for yourself: open up two identical text documents, set one in *Times New Roman* and the other in *Georgia* and compare the reading experience of each.

Response on Twitter to CERN's announcement

Mo Riza @moriza 4 Jul 2012
CERN scientists inexplicably present Higgs boson findings in Comic Sans #CERN #Boson #Comic_Sans

Cosmin TRG @CosminTRG 4 Jul 2012
They used Comic Sans on the Higgs Boson powerpoint presentation... Nope there is no hope for mankind.

HAL 9000 @HAL9000_ 4 Jul 2012
CERN's Higgs presentation just added weight to the theory that Comic Sans is a terrible font.

Colin Eberhardt @ColinEberhardt
Possibly the biggest scientific discovery of our time, the #Higgs Boson, announced in glorious MS Comic Sans Font http://twitpic.com/a3pl0s

fred_SSC @fred_SSC 4 Jul 2012
Dear @CERN: Every time you use Comic Sans on a powerpoint, God kills the Schrödinger's cat; Please think of the cat

Christopher @crgeary 4 Jul 2012
#CERN, Brilliant scientists, terrible at choosing fonts. 'Comic Sans' possibly the worst font choice they could have made.

Nina Lysbakken @NinaLysbakken 4 Jul 2012
OMG! 'Holy geekness, presenting the existence of the #Higgs boson particle using Comic Sans?! #CERN #LHC'

Timmie @longskatedeath 4 Jul 2012
Comic Sans: even if you're a #Cern scientist, you just shouldn't do it. Ever. Nice one #Higgs.

Peter Barna @PeterBarna 4 Jul 2012
Higgs Boson finding is somewhat diminished thanks to the use of the font comic sans. Good job #CERN

Dilia Carolina @DiliaOlivo 4 Jul 2012
#CERN finally finds the #Higgs boson, and Twitter freaks out about the discovery being presented in Comic Sans. lol

Andreas Udo de Haes @AndreasUdo 4 Jul 2012
BREAKING: COMIC SANS IS TRENDING! #higgs #cern #nerds

Renaud's results could be considered subjective as there are many unknown variables that may have influenced the marks. However, he does illustrate the importance of considering both context and readability when choosing a typeface.

Font faux pas

Select the 'wrong' typeface and you can unwittingly commit a font faux pas with the potential to overshadow, or even undermine, the credibility of your message.

In 2012 CERN announced that it had proved the existence of the Higgs boson, or 'God' particle. This was a momentous scientific event but, within an hour of the news, *Comic Sans*, the font in which the announcement was made, was trending higher on Twitter than the discovery itself.[8] It became a major talking point that such an important scientific breakthrough should be announced in a style inspired by comic books. When asked why she chose it, Fabiola Gianotti, the co-ordinator of the CERN program, said 'Because I like it.'[9]

Creating discord

The 'wrong' typeface can be used on purpose to create tension and subvert the meaning of the words. The juxtaposition can prompt the reader to stop and question what they see and this can be used to powerful effect. Director Stanley Kubrick used contrast to powerful effect with music in the film *A Clockwork Orange*, in which scenes of extreme violence are set to classical music, making them feel all the more disturbing.

Designers Why Not Associates created the materials for the 'Apocalypse' exhibition at the Royal Academy of Arts.[10] The title treatment uses *Trade Gothic*, with the counters of the letters punched out to create an unsettling effect, especially when placed over beautiful images.

Alison Carmichael's D&AD award-winning piece C**t[11] features posters with the offensive swear word beautifully hand-lettered and screen-printed in baby pink. The subtitle reads 'Words look much nicer when they're hand lettered' and her poster has appeared in places where saying the word out loud would be completely inappropriate.

LOVE
Klute

Hate
Balmoral

APOCALYPSE
Modified *Trade Gothic*

Brains mistake reading for doing

Studies have shown that our brains can mistake the process of reading for the actual experience it describes. A typeface can convince you that something might be easier to do, a decision might be harder to make, or that a chef is more skilful.

1 Easier to exercise

Psychologists Hyunjin Song and Norbert Schwarz at the University of Michigan researched the idea that our brains can mistake an easy-to-read typeface for the ease of doing something.[12] They tested this by seeing if they could motivate a group of 20-year-old college students to exercise regularly. The students were given written instructions for an exercise routine printed in one of two different typefaces. The easy-to-do instructions were in standard *Arial*, the hard-to-do instructions were in *Mistral*, a paintbrush style typeface that was chosen for being less familiar and harder to read (see page 62).

Those who read the instructions in *Arial* estimated the routine would take 8.2 minutes to complete, compared with the 15.1 minutes estimated by those who read them in the brush style *Mistral*. The 'easy to read' group were more willing to incorporate the regime into their daily routine and the 'hard to read' group thought it would drag on. As Song and Schwarz had anticipated, the participants misread the ease of reading the instructions for the ease of actually doing the exercises.

2 Easier to decide

At the Yale School of Management an experiment was set up in which two groups were asked to read a leaflet with details about a cordless phone.[13] Nathan Novemsky, Associate Professor of Marketing, gave one group the leaflet in an easy-to-read font, and the other group were given it in a difficult-to-read font (see page 62). After reading it they were asked whether they would buy a phone straight away, or if they needed more time to make their decision.

Novemsky predicted that participants would misinterpret the difficulty they had reading the font for difficulty in making a decision and would want more time to make their choice with the hard-to-read font. His predictions were accurate. Only 17% of the participants with the easy-to-read leaflet chose to defer making a choice as opposed to 41% of those with the difficult-to-read leaflet.

Easier to exercise?	Easier to decide?	Skill of the chef?

Exercise Routine
Arial

Exercise Routine
Mistral

'EASY TO READ' FONT
Product features listed in
Times New Roman italic
with no additional effects

'DIFFICULT TO READ' FONT
Product features listed in
Times New Roman italic with
a distracting drop shadow

MENU
Arial bold

Menu
Century italic

3 Easier to play

One of the Type Tasting surveys asks participants to compare four typefaces to music. They are asked to imagine that the shapes of the letters are creating notes and chords as they read, and that these are forming music with rhythm and energy.

When asked which would be the easiest music to play, 62% of the 615 participants chose *Futura*; *Bodoni* was considered the hardest to play.

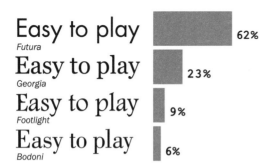

Easy to play
Futura — 62%

Easy to play
Georgia — 23%

Easy to play
Footlight — 9%

Easy to play
Bodoni — 6%

When hard to read is better

If you want people to think that something is easy to do, or maybe that your product is simple to understand or to assemble, then using a typeface that is familiar and easy to read could help you achieve this.

However, if you would like to give the impression that the product requires skill to create then it would be more effective to choose a typeface that requires the reader to put more effort into the reading process. In this instance they could mistake the extra effort needed to read the words for the skill required to create the product. In a second study Song and Schwarz found that participants equated an easy-to-read typeface on a menu with a less skilful chef, but they assumed the chef would be more skilful, and were willing to pay more for the meal, when the menu was set in a more difficult typeface.[14]

An unfamiliar typeface slows our reading down and makes us pay attention, which takes us off autopilot. This can be useful in teaching situations, according to Dr David Lewis, as pupils are 'obliged to invest greater time and attention' in deciphering the words. A high school in Ohio discovered that when students studied from texts in an unfamiliar font, their exam results were higher than those students who had been given the books in a more familiar and readable one. 'Students had to pay closer attention and to think more deeply about what they were reading.'[15]

How many animals of each species did Moses take on the Ark?

This is a question Lewis poses in his book *The Brain Sell: When Science Meets Shopping*[16]. What is your answer? If it is 'two' then you are wrong, because it was Noah who led the animals into the Ark, not Moses.

Lewis explains that the reader is more likely to get a question like this wrong when it is posed in an easy-to-read font. Around 80% got it wrong when the question was put to students in the 1980s. However when a less familiar typeface like *Brush Script* was used then the number who got it wrong went down to around 50%. The unfamiliar type meant that the readers had to pay more attention to reading and in the process they became more aware of what the words actually said.

Proofreading tip

When proofreading a document you have written, try changing it to an unfamiliar or less readable font as this my help you to notice any errors more easily. Remember to change it back when you've finished.

Answers
There are 10 mismatched words:

fast *slow* **gazelle** *heavy* *tortoise* **light** *strong* **weak rat** *elephant*

Typefaces used:

Cooper Black
Palatino Italic

Lawyers typefaces: *Trajan*, *Stencil*, *Cinema* italic, *Copperplate*, *Caslon* italic, *Didot*.

References
1 *The Brain Sell: When Science Meets Shopping* by Dr David Lewis, 2013, Nicholas Brealey Publishing.
2 'Preference Fluency in Choice' by Nathan Novemsky, Ravi Dhar, Norbert Schwarz, and Itamar Simonson, 2007, *Journal of Marketing Research*.
3 'Typographic influences on reading' by Clive Lewis and Peter Walker, 1989, *British Journal of Psychology*.
4 The Nintendo 'quit screen' message is an epigraph that features in T. Michael Martin's *The End Games*.
5 'The meanings of type' by Steven Heller and A. M. Cassandre, 2003, *Eye* magazine.
6 'Hear, All Ye People; Hearken, O Earth (Part 1)' by Errol Morris, 2012, nytimes.com.
7 'The Secret Lives of Fonts' by Phil Renaud, 2006, riotindustries.com.
8 'Higgs Boson Discovery Announcement Made In Comic Sans' by Michael Rundle, 2012, *Huffington Post*.
9 Ibid. Errol Morris.
10 'Design Doyenne: The Answer to London Design is Why Not?' by Margaret Richardson, 2000, creativepro.com, on 'Apocalypse: beauty and horror in contemporary art'.
11 'Alison Carmichael: Exquisite Handjobs', by Gavin Lucas, 2009, *Creative Review*.
12 'If It's Hard to Read, It's Hard to Do: Processing Fluency Affects Effort Prediction and Motivation' by Hyunjin Song and Norbert Schwarz, 2008, University of Michigan.
13 Ibid. Nathan Novemsky.
14 Ibid. Hyunjin Song and Norbert Schwarz.
15 'The Hidden Power of the Font: How fonts make soups tastier, students smarter, and sales easier' by Dr David Lewis, 2014, psychologytoday.com.
16 Ibid. Dr David Lewis.

Online Type Tasting surveys can be found at typetasting.com.

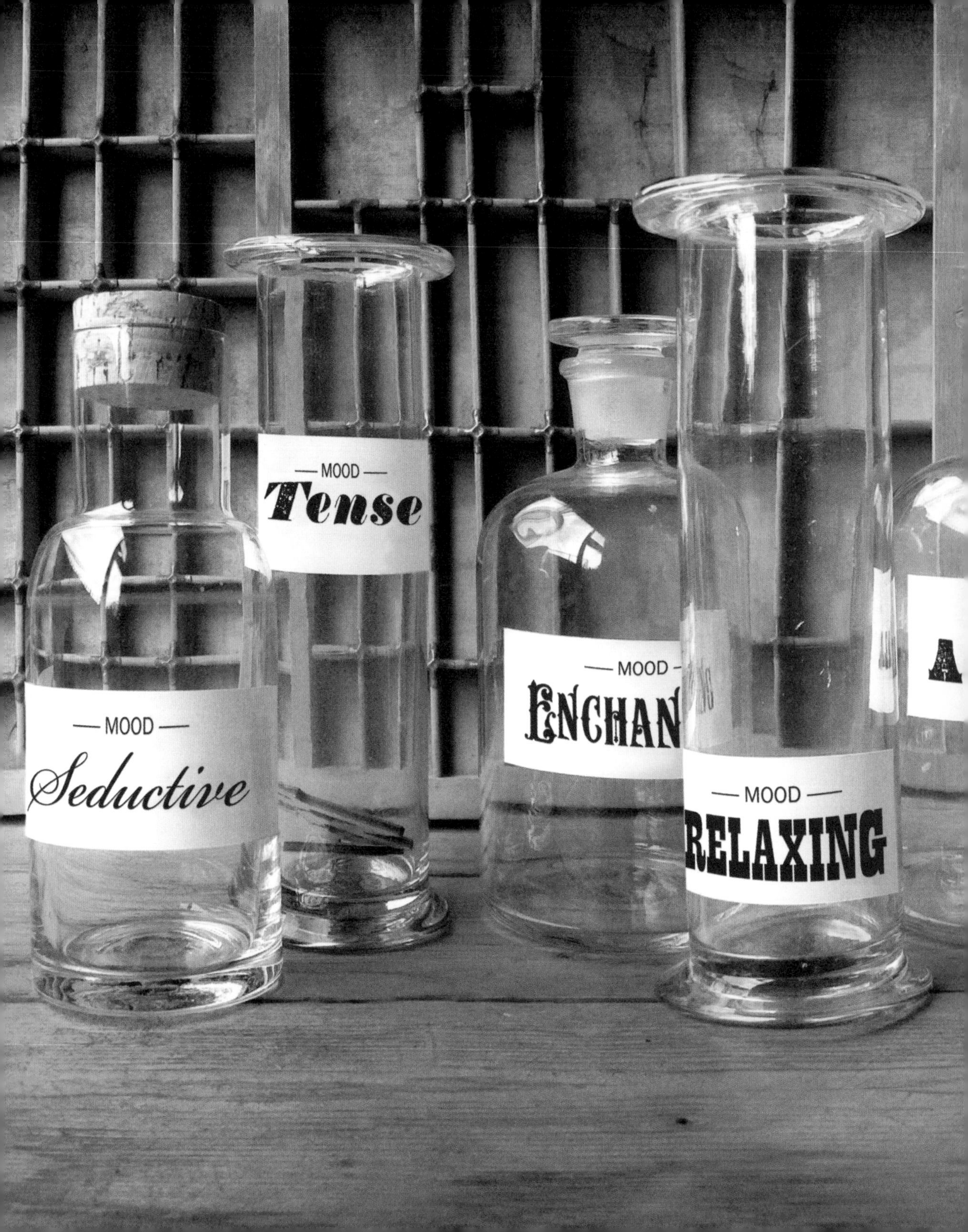
MOOD
Tense
MOOD
Seductive
MOOD
MOOD
RELAXING

7

Setting the mood

Props from a Type Tasting session

1. Which looks the happiest?

happy happy *happy*

☐ ☐ ☐

2. Which is the most excited?

wow *wow* *wow*

☐ ☐ ☐

3. Which is the most angry?

ANGRY ANGRY *ANGRY*

☐ ☐ ☐

4. Which is hiding a secret?

secret secret secret

☐ ☐ ☐

5. Which is the funniest?

haha haha *haha*

☐ ☐ ☐

6. Which is the saddest?

sad sad sad

☐ ☐ ☐

7. Which is the calmest?

calm calm *calm*

☐ ☐ ☐

8. Which is the most deceptive?

deceipt deceipt deceipt

☐ ☐ ☐

9. Which is the most sarcastic?

sarcastic sarcastic sarcastic

☐ ☐ ☐

10. Which is the funniest?

funny funny funny

☐ ☐ ☐

11. Which is the saddest?

sad **sad** *sad*

☐ ☐ ☐

12. Which looks the most scared?

boo! **boo!** *boo!*

☐ ☐ ☐

13. Which is the most tense?

tense tense tense

☐ ☐ ☐

14. Which looks the happiest?

happy **happy** happy

☐ ☐ ☐

Quiz: Emotions

(Turn to page 71 to compare your answers with the results of the Type Tasting online survey).

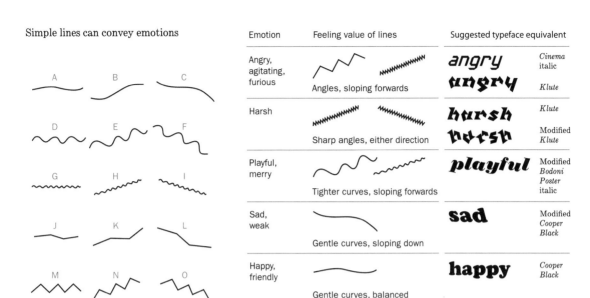

Simple lines can convey emotions

Emotion	Feeling value of lines	Suggested typeface equivalent	
Angry, agitating, furious	Angles, sloping forwards	*angry*	*Cinema italic*
		angry	*Klute*
Harsh	Sharp angles, either direction	**harsh**	*Klute*
		harsh	*Modified Klute*
Playful, merry	Tighter curves, sloping forwards	*playful*	*Modified Bodoni Poster italic*
Sad, weak	Gentle curves, sloping down	**sad**	*Modified Cooper Black*
Happy, friendly	Gentle curves, balanced	**happy**	*Cooper Black*
Calm, gentle	Gentle curves, balanced	calm	*VAG Rounded*

SETTING THE MOOD

In chapter 4 we looked at type as a way to represent both your voice and the breadth of your vocal range. Physically we use our voice, facial expressions, gestures and posture to convey a wide range of emotional clues from the subtle to the dramatic. Typefaces and the way they are used provide a similarly extensive emotional range typographically.

In 1933 Poffenberger and Barrows explored how shapes as simple as lines could communicate emotions.[1] Their theory was that when we look at a line our eyes move along the shape. This turns it into a physical experience that reminds us of the body language we use to express our emotions. They asked participants to match emotions from a list to each of 18 curved and jagged lines sloping in different directions. A line going downwards was shown to make us feel 'doleful', whilst a 'joyous' line takes our eyes upwards.

Their findings are shown in the chart above. This also includes typeface equivalents that I suggest have similar visual qualities to show how the results could relate to type.

Of course, each of our emotions contains a whole range of feelings and these nuances can also be conveyed typographically. For example here are six types of 'happy'.

Exuberantly **happy** Whistfully **happy** Faking it *happy*

Satisfyingly **happy** Calmly happy Agressively **happy**

Satirical Serifs

Psychologists Samuel Juni and Julie Gross asked 102 New York University students to read a satirical article from the *New York Times*. Each was given the reading randomly printed in either *Times New Roman* or *Arial*.

satirical satirical
Times New Roman   *Arial*

Afterwards they were asked to rate their response to what they had read. They rated the article as being funnier and angrier, in other words more satirical, when it was read in *Times New Roman*.[2]

emperatures and the general feel

ılking point early in the week beco

ɔol with plenty of showers but also

ɔells. The brisk winds will graduall'

ressure centre that's driving our w

(a) *VAG Rounded*

One of the questions on page 66 is which of the three options is the most sarcastic. In the online version of the survey *Helvetica* was rated the least sarcastic and *Footlight* italic the most sarcastic by a considerable margin.

sarcastic
Footlight italic

sarcastic
Georgia

sarcastic
Helvetica

62%

23%

15%

Crowd control

'Typography has a visceral and direct effect on everybody who reads.' Martin McClellan[3]

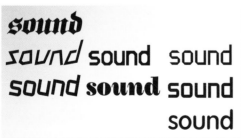

Anxious

Calm

Active   Static

If typeset words were sounds, how would they make you feel? After asking a wide range of people this question it appeared that the results could be broken down into two categories: the shape linked to emotions, and the axis (or slope) of the letters linked to energy. Curve-shaped letters are considered calming, whilst angular-shaped letters evoke anxiety. Balanced, non-sloping letters appear static, whilst angled letters appear active.

I find it interesting that type and graphic design can evoke a physical response in the viewer. In their study 'The aesthetics of reading',[4] psychologists Kevin Larson and Rosalind Picard measured how people express their discomfort at reading poorly designed, difficult-to-read materials through facial expressions like frowning.

When I show typefaces on screen during workshops I sometimes see the audience react, often without realising they are doing so. Recently I showed the two typefaces at the top of these pages in succession and saw people flinch and turn away from the screen when I showed slide (b).

The typefaces above have been chosen to provoke a reaction dramatic enough for you to notice, to show how type could alter mood. Could typefaces be used to energise a classroom of unruly children, or to calm an agitated crowd?

emperatures and the ge
rill be a talking point ear
e windy and cool with pl
rier, brighter spells. The
use as the low pressure

(b) *Klute*

Comfort reading

Could you use fonts to alter your own mood? In theory you could select a font to energise you when you hit the four o'clock slump, or to be increasingly relaxing as you read before going to sleep at night.

Act now
Wake up
Laugh
Think
Calm down
Go to sleep

Klute, Helvetica italic, Footlight italic, Caslon, VAG Rounded, Helvetica in different weights.

Threat response

The reason why we react negatively to angular letter forms is because we are programmed to respond to these shapes; recognising danger has been crucial to human survival. The area of the brain where fear is processed is the amygdala and this plays a key role in alerting us to potential threat.[5] The amygdala is triggered by facial expressions of threatening emotions like anger and fear, and by sharp and jagged shapes. This causes us to feel fear and be alert to possible danger. Non-threatening facial expressions and rounded shapes do not activate the fear response and so we experience them as safe and friendly.

There is a parallel between what we experience in the physical world and how this influences our interpretation of typeface shapes. Type can be seen as mirroring the emotions we display in the real world through our facial expressions and gestures. When we are happy our faces become round with a wide smile and our body language is open. By contrast, an angry frown expression is pinched and angular and an attacking animal is all jagged teeth and claws.

Type also mirrors the way your handwriting communicates your mood or emotions. When writing quickly your writing is italicised and when angry it becomes bold and deliberate.

The parallel between the physical world and the shape of type

Friendly, happy and calming

Curves, soft shapes

Friendly

Balanced and geometric shapes

VAG Rounded

Unfriendly or impersonal

Angular shapes, sharp corners

Unfriendly

Klute

Open vs restrained

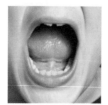

'O' shape

Open   o

Large & round letters

Restrained   o

Small, narrow or condensed letters

Avant Garde / Franklin Gothic

Dynamic

Directional

↑

Dynamic

Times New Roman italic

Professional vs informal

Formality, weight and contrast

a **a**

Professional

Formal, moderate weight and contrast

Informal

Exaggerated weight and contrast

Caslon / Cooper Black

Easy vs interesting

Shape

S **S**

Easy   s

Simple shapes

Interest s -ing

Complex shapes

VAG Rounded / Bodoni Poster italic

Traditional

Serifs

⊥ ⊥

Intellectual

Hand drawn feel, curved serifs

TIMELESS

Roman inscriptions, curved serifs

Garamond / Trajan

Modern

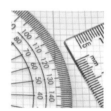

Sans serifs and clean lines

○ ⊥

Modern

Geometric shapes

Modern

Angled corners and fine serifs

Futura / Didot

1. Which looks the happiest?			2. Which is the most excited?			3. Which is the most angry?		
happy	happy	**happy**	wow	wow	wow	ANGRY	ANGRY	**ANGRY**
34%	25%	**41%**	**63%**	12%	25%	21%	37%	**42%**

4. Which is hiding a secret?			5. Which is the funniest?			6. Which is the saddest?		
secret	secret	secret	haha	haha	**haha**	sad	sad	sad
16%	26%	**58%**	29%	15%	**56%**	25%	**39%**	36%

7. Which is the calmest?			8. Which is the most deceptive?			9. Which is the most sarcastic?		
calm	**calm**	calm	deceipt	deceipt	deceipt	sarcastic	sarcastic	sarcastic
6%	**86%**	8%	14%	23%	**63%**	15%	23%	**62%**

10. Which is the funniest?			11. Which is the saddest?			12. Which looks the most scared?		
funny	funny	funny	**sad**	sad	sad	boo!	boo!	**boo!**
22%	22%	**56%**	**47%**	19%	34%	16%	29%	**55%**

13. Which is the most tense?			14. Which looks the happiest?		
tense	tense	**tense**	happy	happy	happy
34%	31%	**35%**	**42%**	32%	26%

1 *Cooper Black.* 2 *Bodoni Poster* italic. 3 *Futura.* 4 *Futura, Georgia, Footlight.* 5 *VAG Rounded, Mateo, Candice.*
6 *Futura, Georgia, Footlight.* 7 *VAG Rounded.* 8 *Futura, Georgia, Footlight.* 9 *Futura, Georgia, Footlight.*
10 *Futura, Georgia, Footlight.* 11 *Cooper Black.* 12 *Cooper Black.* 13 *Cinema Gothic.* 14 *Futura, Georgia, Footlight.*

Results are based on 1,922 participants; the majority are non-designers from the UK and the US.

References
1 'The Feeling Value of Lines' by A. T. Poffenberger and
B. E. Barrows, Psychological Laboratory, 1933, Columbia
University.
2 'Emotional and persuasive perception of fonts'
by Samuel Juni and Julie S. Gross, 2008.
3 'Gutenberg and how typography is like music'
by Martin McClellan, mcsweeneys.net.
4 'The aesthetics of reading' by Kevin Larson (Microsoft)
& Rosalind Picard (MIT), 2007.
5 'Functional connectivity between amygdala and facial
regions involved in recognition of facial threat' by Motohide
Miyahara, Tokiko Harada, Ted Ruffman, Norihiro Sadato,
Tetsuya Iidaka, 2011, *Oxford Journals.*

Online Type Tasting surveys can be found at typetasting.com.

8

Type is a time machine

Image courtesy of The Museum of Brands

TYPE IS A TIME MACHINE

Type is interwoven into the rituals of our lives through the packaging on the products that we use every day. At the time we may pay little attention as the logos and product names are so familiar that we reach for them almost unconsciously. But when we come across old packaging years later we can be transported back in time and find ourselves immersed in a memory. These memories can start interesting conversations that give us an insight into the lives people have lived.

My dad grew up in Maghera, Northern Ireland, and when he was a teenager in 1949 his father opened the Hyndman's bakery. My friends recently went there and they brought back a pack of Hyndman's potato bread for me. I photographed the label to show to Dad, who rang me as soon as he saw it and spent an hour happily reminiscing about his youth, when he worked in the bakery after school and when he came home from college in the summer holidays.

He shared an array of memories from years before I was born that were all triggered by this simple label. These were stories I would otherwise not have heard, because they were not the remarkable tales of major events and adventures, instead they were gentle memories of his everyday life. Hearing them gave me a real glimpse of my dad as a young man, which melted away the years.

Reminiscent packs

The Museum of Brands in west London contains a treasure trove of memories.[1] It houses a collection of packaging which dates from the Victorian era to the current day and reveals the way we have lived and shopped decade by decade.

Following work with organisations like Arts 4 Dementia, the museum's team has created two themed packs for dementia patients that utilise well-known household brands from the past century. These are used to evoke memories and

Madeleine cakes, 'Let's Go Shopping' pack courtesy of The Museum of Brands, Hyndman's potato bread

encourage conversations. The museum has found that these have had a unanimously positive response and care workers have found using them to be a very successful way to provoke discussion.

Encouraging conversation is considered important by organisations working with the elderly. Hannah Stewart, an Adult Care Commissioning Officer, says 'focusing on positive memories from the past can be comforting for a person with a condition which can cause anxiety, agitation and short-term memory loss. A person with dementia can often talk about the past with confidence and certainty, giving them a sense of control and revealing their unique personality to others.'

The sweets section in the museum is organised by decade, so you can guess a person's age from the cabinet they stop and stand in front of. The selection of 1970s sweets gave me a rush of childish excitement when I first saw it, bringing back the memory of my grandparents' house at Christmas, the fresh tobacco of my grandad's

cigarettes and the smell of Christmas lunch cooking. I remembered giggling with my brother and sister, getting increasingly 'sugar drunk' as we tore open wrapping paper and ate the sweets we found inside. These 1970s sweets were the backdrop to Christmases of my childhood and it was quite magical to find myself transported back.

Involuntary memory

In his book *À la Recherche du Temps Perdu*[2] French novelist Marcel Proust famously talks about this type of unexpected nostalgia, which is experienced without any conscious effort, as involuntary memory. He describes the episode of the madeleine in which he, as an adult, eats a madeleine cake dipped in tea and experiences an exquisite pleasure. The taste and the smell evoke sensory memories stored in his brain that begin with the emotions. These are followed afterwards by the memory of the actual event (his aunt feeding him madeleines dipped in tea when he was a child).

Zeitgeist

A typeface captures the spirit of when it was designed and is a permanent record of that moment in time. In this way typefaces document social history and chart developments in technology. They reflect our wide-ranging tastes and aesthetics, from highbrow to popular culture.

The typeface *Jenson* is based on Nicholas Jenson's Venetian Old Style typefaces of the 1470s, created in 1996 by Robert Slimbach. Using the font today provides a direct link to the early history of type.

Jenson

In the 1820s Louis John Pouchée created a series of highly decorative typefaces.[3] These were carved from wood and adorned with images of the era ranging from farmyard scenes to Masonic symbols, giving a glimpse into the lives and pastimes of the 19th century.

Pouchée letters courtesy of St Bride Library & Archives.

Chicago

Susan Kare originally designed the typeface *Chicago* for the Apple computer interface in 1983.[4] It became an iconic part of Apple's brand identity and today it encapsulates the technological revolution that transformed the design industry and launched a new era of personal computing.

Looking at this font takes me back to my first days as a graphic designer and the awe we all felt at the possibilities offered by these tiny computers, despite them having only 128Kb of memory.

Some type styles can come back into fashion, or be appropriated by a new generation. Art Nouveau was popular in Paris in the 1890s and it remains on café signs around the city and on the Paris *Métro*. The style regained popularity in the late 1960s when it inspired the letter styles of the psychedelic art movement. The example shown below is *Victor Moscoso* and is based on the lettering of the 1960s artist of the same name.

Psychedelic Art

Instant nostalgia

Type can transport you to an imagined nostalgia that you may not have experienced first-hand, but which has become real to you through the experience of film and television. *Grease*, *Back to the Future* and *Mad Men* have recreated the 1950s so vividly for me that I feel I 'remember' the decade even though it was before I was born.

A photo can be altered to look like it was taken decades, or even centuries, ago by using a filter effect like a 1970s Polaroid or by changing it to sepia. The page opposite shows how we can do the same thing with words by applying different fonts; a word can be transported to different historic eras to create an 'instant nostalgia' look.

Fonts are like Instagram filters for words

1400s
Medieval

1800s
Mexico

1960s
Psychedelia

1970s
Army adventure

1700s
Opulence

3000s
Sci-fi

Late 1800s
Precious

1930s
Art Deco

1800s
Wild West

2000s
Geometric

Late 1800s
Victorian

1920s
Bauhaus

1900s
Circus

Late 1800s
Ornate Victorian

1980s
High-tech

Late 1800s
Art Nouveau

References
1 The Museum of Brands, London. museumofbrands.com.
2 À la recherche du temps perdu by Marcel Proust, 1913.
3 'Pouchée's lost alphabets' by Mike Daines, 1994, *Eye*.
4 'A history of Apple's typography' by Walter Deleon, 2013, technoblimp.com.

9

Fonts give words a personality

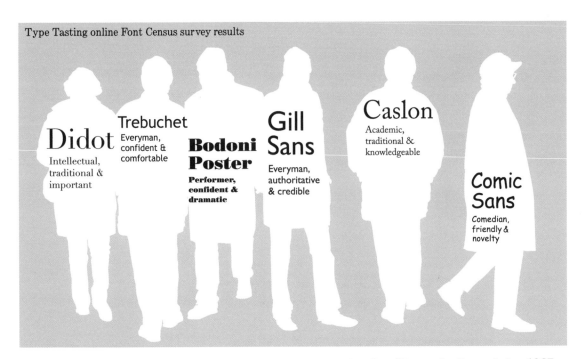

Didot
Intellectual, traditional & important

Trebuchet
Everyman, confident & comfortable

Bodoni Poster
Performer, confident & dramatic

Gill Sans
Everyman, authoritative & credible

Caslon
Academic, traditional & knowledgeable

Comic Sans
Comedian, friendly & novelty

FONTS GIVE WORDS A PERSONALITY

Helvetica, Times New Roman and *Comic Sans* walk into a bar. The barman turns to *Comic Sans* and says, 'Sorry we don't serve your type in here'.

This familiar joke demonstrates that fonts have personalities that we recognise easily. Some are unassuming, whilst others are larger-than-life like Mike Lacher's version of *Comic Sans*[1], who says, 'People love me. Why? Because I'm fun. I'm the life of the party', then goes out and gets drunk with *Papyrus*.

Caslon Fraktur
Garamond
Bodoni

The most significant designs from the early centuries of type have become intrinsically linked with their creators and countries of origin. In some instances it is considered that they have come to

represent nationalities and cultures. In her 1987 essays on modern typography, Jadette Laliberté[2,3] matches *Fraktur* with Germany, *Garamond* with France, *Bodoni* with Italy and *Caslon* with England. However, she explains that with the 'new typography' movement of the 1920s, designers began to reject these typefaces in favour of new ones that were free from historical and cultural 'baggage'.

In his 1991 book *Graphic Arts and Book Design*[4] influential designer Jan Tschichold also talks of typefaces having unique personalities and stresses that they must be chosen appropriately to match the character of the content.

The personality of the typeface sets the tone for what you're about to read in the same way that your first impressions of how a person looks influences how you will listen and respond to them.

Font Census

As a graphic designer I have an awareness of typeface personalities which I use in my work. However, I realise that my learned knowledge (see page 45) comes mostly through observation and intuition, but not from actual data. As part of my

Times New Roman
Intellectual, confident & neutral

Helvetica
Everyman, conventional & neutral

Garamond Italic
Artist, confident & classic

Baskerville
Intellectual, traditional & neutral

Friz Quadrata
Leader, honest & classic

Lubalin Graph
Idealist, confident & classic

research I wanted to test out my assumptions and asked myself the following questions:

(a) Can I prove that typefaces have uniquely different personalities?

(b) Is there significant agreement from others on the different personality types?

(c) Do the opinions of non-designers and designers differ in any way?

I created an online Font Census[5] that began with 25 regularly used typefaces. Survey participants are randomly assigned an anonymous typeface and asked to rate various aspects of its personality along with a series of more descriptive questions like, 'What job would it do?'

The results show that (a) typefaces do have uniquely different personalities, and that (b) there is significant agreement on the personality types. There are detailed breakdowns of three examples on page 132, and the illustrations and charts throughout this chapter show the results. At the time of writing each of the typefaces has received 250–350 responses, the majority of these are from non-designers in the UK and the US.

Designers and non-designers

'Graphic designers can get a bit serious and beard-stroking about typefaces.' Angus Montgomery, editor of *Design Week*

In the surveys, participants are asked whether they have professional experience working with type. On analysing the responses it became clear that type professionals (designers) can attribute more qualities to typefaces and describe them using more complex language. They also show more deference for typefaces with historical status such as *Helvetica*, *Gill Sans* and *Bauhaus*.

Helvetica

Non-designers: everyman, meh, dull.
Designers: intellectual, intelligent, stylish.

Gill Sans

Non-designers: hmm…, neutral but kinda quirky, boring, ordinary, readable, official, honest, office.
Designers: simple and elegant, classic design, good, solid British Modernist typeface, lots of authority, not particularly exciting, clean, crisp, classic.

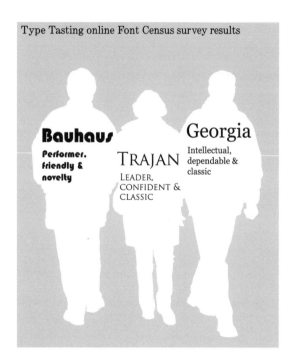

Bauhaus
Performer, friendly & novelty

TRAJAN
LEADER, CONFIDENT & CLASSIC

Georgia
Intellectual, dependable & classic

Of course, if you recognise a font's personality then it stands to reason that you can make a pretty good guess at what it might like to drink.

Helvetica, *Times New Roman* and *Comic Sans* walk into a bar. *Helvetica* 'the everyman' orders a pint of lager, *Times New Roman* 'the intellectual' selects a respectable white wine and *Comic Sans* 'the comedian' gets the shots in.

Helvetica

Times New Roman

Comic Sans

Bauhaus

Non-designers: silly, a clown, friendly, doughnuts.
Designers: architecture, art movement, technical.

I was surprised when descriptions like silly and a clown first came in for *Bauhaus*. This is a typeface associated with the influential art school of the same name, of Herbert Bayer's universal alphabet experiments and an ideology that inspired Modernist design and architecture. It seems that the inclusion of *Bauhaus* in the default Microsoft font menu has separated it from its history, leaving it to be judged purely on its stylised shapes.

The answers demonstrate that those of us who work with type have accumulated a wealth of knowledge about typefaces from working with them every day. We know their history, origins, often who designed them, we have a mental catalogue of associations to refer to and we notice the subtle differences between styles. Professor Eva Brumberger refers to professionals typically attributing 'more qualities to a given typeface'.[6]

Type Tasting online Font Census survey results
Identification of personality attributes

Personality (bold)
Values (regular)
Style (italic)

The highest rated attribute is listed first.

Baskerville

Intellectual, academic, wise
Traditional, conventional, trustworthy
Neutral, credible, knowledgeable

Bauhaus

Performer, artist, jester
Friendly, cheerful, imaginative
Novelty, arty, modern

Bodoni Poster

Performer, leader, idealist
Confident, ambitious, cheerful
Dramatic, classic, arty

Braggadocio

Performer, artist, jester

Obnoxious, imaginative, confident

Dramatic, flamboyant, arty

Caslon

Academic, intellectual, informed

Traditional, trustworthy, old-fashioned

Knowledgeable, neutral, important

Cinema Italic

Commander, doer, thinker

Logical, self-controlled, capable

Machine-made, modern, practical

Clarendon

Doer, leader, thinker

Traditional, professional, confident

Practical, traditional, classic

Cocon

Performer, artist, jester

Friendly, cheerful, relaxed

Comfortable, quick 'n' easy, feminine

Comic Sans

Comedian, everyman, storyteller

Friendly, welcoming, loud

Novelty, quick 'n' easy, comfortable

COPPERPLATE

Leader, intellectual, thinker

Dignified, authoritative, confident

Classic, formal, polished

Didot

Intellectual, thinker, leader

Traditional, confident, professional

Important, sophisticated, polished

Franklin Gothic

Everyman, leader, doer

Dependable, capable, confident

Practical, classic, comfortable

Friz Quadrata

Leader, thinker, intellectual

Honest, confident, professional

Classic, comfortable, distinctive

Futura

Leader, idealist, thinker

Modern, confident, capable

Neutral, practical, comfortable

Garamond

Intellectual, thinker, leader

Responsible, professional, well informed

Classic, formal, traditional

Garamond italic

Artist, performer, idealist

Confident, cheerful, imaginative

Classic, arty, feminine

Georgia

Intellectual, thinker, leader

Dependable, confident, professional

Classic, practical, safe

Gill Sans

Everyman, doer, idealist

Authoritative, capable, dependable

Credible, practical, safe

Helvetica

Everyman, doer, idealist

Conventional, confident, modern

Neutral, credible, calm

Lubalin Graph

Idealist, doer, leader

Confident, professional, responsible

Classic, practical, traditional

Monotype Corsiva

Artist, idealist, performer

Friendly, dignified, pretentious

Elegant, feminine, classic

Times New Roman

Intellectual, academic, informed

Confident, professional, honest

Neutral, traditional, classic

TRAJAN

Leader, thinker, intellectual

Confident, authoritative, traditional

Classic, traditional, formal

Trebuchet

Everyman, idealist, doer

Confident, friendly, honest

Comfortable, modern, practical

VAG Rounded

Everyman, idealist, doer

Friendly, cheerful, relaxed

Comfortable, quick 'n' easy, practical

Garamond
Brown leather brogues

Garamond italic
Kitten heels

Franklin Gothic
Utilitarian work shoes

Georgia
Comfortable loafers

Monotype Corsiva
Ballet flats

Bauhaus
Sparkly pumps

Friz Quadrata
Bright trainers

Gill Sans
Leather-lace ups

Trajan
CLASSIC DRESS SHOES

Lubalin Graph
Doc Martins

Braggadocio
Fashion wedges

Helvetica
Sensible work shoes

Didot
Stilettos

Futura
Manolos

Trebuchet
Cheap sneakers

Clarendon
Walking boots

Type Tasting online Font Census survey results 'What job would a font do?'

Bauhaus

Write comedy articles, comic bookshop clerk, fast food industry, something creative, mad scientist, demonstrate gadgets at the state fair, cheerleader, bartender, entertainer, comedian, talk to children or teens, light-hearted newsreader, record shop employee, jazz singer, fashion blogger, nightclub owner, bagel shop owner, tell jokes, clothes shop, a fun job, New York stylist, cabaret lounge host ...

Bodoni Poster

Impresario, fashion designer, photographer, writer, gossip columnist, journalist, checkout clerk, make-up artist, stylist, sell fashion, sell lollipops, fashion advisor, promo girl, beautician, graphic arts, work for a tabloid, business, maybe fashion, storyteller, paparazzi, car salesman, web designer ...

Braggadocio

Film poster designer, art gallery coordinator, truck driver, circus performer, advertise financial services for artists, radio host, out of work actor, hawker, make sandwiches, sensation journalist, historian, art deco historian, something retro, construction, bar mixologist, showman, taste tester, sell vintage clothes, theatre usher, hipster, speakeasy bouncer, set designer for broadway plays ...

Cinema Italic

Maths teacher, IT, computer programmer, engineer, robot, data analysis, technology, coder, astronaut, some dull data processing job in a dimly lit office, meteorologist, nanotechnologist, audio engineer, sci-fi author, rocket scientist, repair computers, scientist, auto parts sales, makes annoying tele-marketing calls, artificial intelligence engineer, conspiracy theorist, new technology journalist ...

Clarendon

Design professional, write sub headings, inform people, newspaper editor, professor, teacher, stenographer, reporter, make road signs, write headlines, paper boy, infographics designer, tell the news, advertising, book shop manager, publicist, sign maker, proofreader, postman, storyteller, party planner, social news journalist ...

Cocon

Music reviewer, baker, daycare, coach, comic art, public relations, shopping mall sales clerk, yoghurt shop employee, ice cream scooper, retail, barista, dancer, teach young children, dog walker, gossip, dentist, skater, ditzy receptionist, toy seller, cook, cocktail bartender, ice cream parlour, people person, baker, entertainer, ad agency, manicurist, dog groomer, psychologist, carnival worker, event planning, journalist ...

COPPERPLATE

Professor, accountant, an owner of a business who is old and grouchy and doesn't like change, bartender, classifieds, mortician, banker, statistics, analyst, financial, accountant, guide, doctor, stockbroker, history teacher, librarian, corporate lawyer, finance, private investigator, business tycoon, monarch, communications, cellist, financing, bar owner ...

Didot

Fashion journalist, finance, host, academic, critic, product designer, advertising, architect, lawyer, write personal ads, attorney, concierge, actuary, fashion blogger, banker, advertiser, fashion magazine editor, lawyer, fashion stylist, decorator, something intriguing, hairdresser, English professor, publishing, female CEO, fashion designer, media consultant, artist, investment banking, publicist, magazine editor ...

Franklin Gothic

Low-level bureaucrat, customer service, tell you information, journalist, nurse, bureaucrat, high-end desk job, coach, fireman, cashier, communicate information, clerical, magazine art director, design instructions, newsreader, secretary, human resources, educator, something practical, print newspapers, boring office job, reporter, professional introductions, local news journalist, hotel manager...

Friz Quadrata

Janitor, relay current events, analyst, photographer, creative, bank loan officer, dry cleaning, event planner, fried chicken shop cook, researcher, storyteller, customer assistance, museum exhibit curator, bartender, human resources, banker, look important, fast food industry, design invitations, public speaker, cyclist, pharmacist, party planner, lawyer but not the cool kind, travel guide...

Futura

Web designer, journalist, host or hostess at a mid-range restaurant, designer, nutritionist, sign maker, advertising, director, engineer, fashion designer, advertising copywriting, sales or marketing president, blogger, professional, write manuals, painter, stock broker, announcer, upper level business, marketing, small business owner, type expert, graphic designer, instructor, printer, assistant, government official, entrepreneur, motivator, cool young entrepreneur...

Garamond

Journalist, councillor, lawyer, teacher, accountant, ombudsman, politics, judge, finance, small business investor, chief executive, attorney, historian, stockbroker, head teacher, tell the facts, editor, informed

journalist, educator, diplomat, business person, editor, novelist, librarian, public servant, clerk, English professor, government, doctor, banker, manager, traditional journalist...

Garamond italic

Celebrity news, television comedian, fashion stylist, personal assistant, sell dresses, artist, survey women in a magazine quiz, plan country club galas, nail painter, singer, liberal arts, interior decorator, sell make-up or clothes, hostess, pastry chef, wedding planner, cook, butler, expert on love,sell you an expensive purse, barista, hostess, look pretty, gardener, beautician...

Georgia

Collect coupons, secretary, researcher, magazine editor, retail sales, writer, journalist, bookshop owner, human resources, customer service representative, call centre, librarian, inform, business analyst, typist, editor, university professor, basic, features writer, office worker, informer, convey news, blogger, fact checker...

Gill Sans

Newsreader, bank teller, journalist, receptionist, work for the BBC, bookkeeper, middle management, cook, announcer on the London Underground, write for a newspaper, teacher, office clerk, interpreter, insurance adjuster, poet, talk to people, doctor, publicist, work at a gym, sit at a desk, write instruction manuals, explain things, something neutral, secretary, middle class job...

Helvetica

Paperwork, give information, customer service, voice of

reason, graphic designer, write serious headlines, advertising, civil servant, deliver news, used car salesman, social media, deliver up-to-date news, something in a cubical, make things easier for people to read, count beans, police, make signage, instructor, census-taker, broker, desk job, insurance, computer scientist, create instructions, dentist, taxi driver...

Lubalin Graph

Art shop employee, graphic designer, flirting tips for men columnist, impresario, shelving installer, new innovation journalist, customer service complaint desk, hula hooper, middle manager, shoe salesman, announcer, journalist, city news reporter, stylist, researcher, make-up artist at a department store, design official government notices, airline attendant, infographic designer, work at a call centre, communicate cutting edge tech research, write how to articles...

Monotype Corsiva

Homemaker, bake pastries, event planner, interior designer, greeter, astrologist, concierge, craft or food blogger, wedding planner, Hallmark card maker, sell shoes to women, cosmetics, waitress, artist, a duke, beautician, stylist, gossip columnist, author of fairy tales, hostess, queen or king, calligrapher, judge, baker, celebrity, give beauty tips, florist, sell perfume, invitation designer...

TRAJAN

Secretary with a quill, Oscar-winning film director, lawyer, business, banker, attorney, accountant, announcer, doctor, senator, door-to-door

sales, author, event planner, matchmaker, Roman senate note-taker, advertiser for gold and precious metal, Roman emperor, write letters and professional papers, archeologist, write small print, professor, doctor, bureaucrat, entrepreneur, president, stockbroker...

Trebuchet

Provide technical information, teacher, bore you with hyperbolic headlines, information giver, entry level graphic design, greeter, cashier, daycare, mailman, waitress, retail clerk, secretary, provide easy reading, communications, insurance sales, lifeguard, office worker, government worker, computer programmer, telemarketing, administrative assistant, social worker...

VAG Rounded

A hippy with no job, give beauty advice, greeter, preschool teacher, letters to the editor columnist, mow lawns for neighbours, elementary teacher, bus driver, surfer, carhop, primary school teacher, office admin, answer phones, student, art teacher, do simple office jobs, volunteer, cartoonist, happy news reader, put people to sleep, babysitter...

Personality categories

In her 2003 study 'The persona of typeface and text'[7], Professor Eva Brumberger found that typefaces 'divide cleanly' into three distinct categories, with no overlaps.

Elegance	Commercial Script*
Elegance	Apple Chancery*
Elegance	Casablanca Antique
Elegance	Harrington
Friendliness	Bauhaus
Friendliness	Comic Sans
Friendliness	Van Dijk
Friendliness	Lucida Sans italic
Directness	Arial
Directness	Garamond
Directness	Times New Roman
Directness	Square 721

When you look at the groupings, the 'elegant' typefaces all have hand-drawn or calligraphic qualities. The 'friendly' typefaces all have rounded shapes and, if hand-drawn, are created with less skill than their 'elegant' counterparts. The 'direct' typefaces are traditional and more neutral Roman or Sans Serif text typefaces.

*Similar typefaces have been substituted. *Commercial Script* has been used in place of *Counselor Script*, *Apple Chancery* in place of *Black Chancery*.

Five letterforms

Jo Mackiewicz compared the five letterforms 'J', 'a', 'g', 'e' and 'n' from different typefaces to analyse the physical characteristics that differentiate between 'friendly' and 'professional' typefaces.[8] Note: not all attributes need to be present in a typeface for it to fit the classification; for example, a font may still appear 'friendly' if it does not have an oblique crossbar on the 'e'.

Friendly: Sans Serif or Serif typefaces

1 Rounded terminals
2 Oblique crossbar on the 'e' letterform
3 Single-storey 'a' letterform
4 Single-storey 'g' letterform
   Examples: *VAG Rounded*, *Frankfurter* ('e')

Professional: Sans Serif

5 Moderate weight
6 Moderate contrast between thick and thin
7 Moderate x-height to cap-height ratio
8 Horizontal crossbar on the 'e' letterform
9 Double-storey 'a' letterform
   Example: *Helvetica*

Professional: Serif

10 Moderate weight
11 Moderate x-height to cap-height ratio
12 Horizontal crossbar on the 'e' letterform
13 Double-storey 'a' letterform
14 Double-storey 'g' letterform
   Example: *Times New Roman*

Baskerville

Imparts dignity to the true old face character giving that added note of brilliance.

Modern No. 20

Suitable for forceful advertisement display and for all official and commercial printing where severity is the essential requirement.

2

'HEAVY MACHINERY NOT FLUFF FLUFF'

3 **The adaption of lettering to advertising moods.**

Dainty ROMAN *Italics*
For grace, elegance and feminine appeal.

CLASSIC ROMAN
For conservative dignity, permanence and beauty.

Texts Church Gothic UNCIAL
For antiquity, quality, craftsmanship and reverence.

GOTHIC & BLOCK
For strength, power and a sturdy atmosphere.

University Roman, Goudy Ornate, Caslon italic, Trajan, Cloister Black, Duc de Berry, Tomism, Futura bold, Rockwell extra bold.

Instructions for jobbing printers

1 Type sampler catalogue 1924[9]
Historically, type foundries produced and distributed metal and wooden type to printers who selected the fonts (typeface styles and sizes) from a type sampler catalogue. These were written in the everyday language of a jobbing craftsman and not the intellectual language of typographic academia. The usage suggestions on the left are from a 1924 catalogue by Stephenson, Blake & Co.

2 100 Type-Face Alphabets 1974[10]
Author Jacob Biegeleisen recommends that the Slab Serif *Stymie* bold italic should be used 'in the fields of industry, men's wear, heavy machinery, electronics etc, rather than the flimsy feminine precincts of lingerie, cosmetics or other such fluff fluff'. (*Rockwell* bold italic has been used in its place in the illustration on the left).

3 Speedball Text Book 1915[11]
Speedball books by Ross F. George gave practical advice on lettering and poster art for both amateur and professional calligraphers and signwriters. 'The adaption of lettering to advertising moods' on the left gives instructions on how type can be adapted to create different moods ranging from 'grace' and 'elegance' to 'a sturdy atmosphere'.

References
1 'I'm Comic Sans, Asshole' by Mike Lacher, mcsweeneys.net.
2 'The rhetoric of typography: The persona of typeface and text' by Eva R. Brumberger.
3 'La typographie moderne: conséquence de la révolution industrielle?' by Jadette Laliberté, 1987, *Communication et Langages.*
4 'Graphic arts and book design' by Jan Tschichold, 1996, *The form of the book: Essays on the morality of good design.*
5 Type Tasting Font Census can be found at typetasting.com/fontcensus.html.
6 'The rhetoric of typography: The Awareness and Impact of Typeface Appropriateness' by Eva R. Brumberger, 2003, *Technical Communication.*
7 Ibid. Eva R. Brumberger.
8 'How to use five letterforms to gauge a typeface's personality: a research-driven method' by Jo Mackiewicz, 2005, University of Minnesota Duluth.
9 From the collection of the St Bride Library, London.
10 *Book of 100 Type-Face Alphabets* by Jacob I. Biegeleisen, 1974.
11 *Speedball 14th Edition: Speedball Text Book. Lettering, Poster Design, for Pen or Brush* by Ross F. George, 1941.

Online Type Tasting surveys can be found at typetasting.com.

10

Fonts reveal YOUR personality

Zoltar's Font Fortune booth at a Type Tasting event

Fonts are clothes for your words to wear

Saturday night out

Relaxing Sunday

Monday at work

Elegant French Script

Sophisticated Didot

Formal Futura

Casual Amadeo script

Everyday Helvetica

Comfortable Cooper Black

Professional Garamond

Credible Gill Sans

Confident Clarendon

FONTS REVEAL YOUR PERSONALITY

Fonts are like typographic selfies. You are drawn to typefaces that reflect your values and aesthetics, and you dislike them when they do not. Your loyalty can be called into question if a brand you pay allegiance to changes its logo typeface to one you no longer identify with. This is illustrated by the outcry from Gap customers when the logo font was changed to *Helvetica* bold, which they thought looked 'cheapy, tacky, ordinary' (page 16).

You interact with fonts through the brands you surround yourself with. You identify with the values of the newspaper you choose to read, the brands you put in your shopping basket, the clothes you wear and the car you drive. As discussed in chapter 1, you are curating the fonts that surround you in your everyday life through the logos on the products you buy and collect.

Designers understand how important typefaces are when creating brand identities, and that a successful one can become highly recognisable from its lettering alone. You know from a single letter whether a cola is Coca-Cola or Pepsi, a TV channel is BBC or ITV, a browser is Yahoo or Google.

Non-verbal communication

When somebody is speaking we take less than 10% of the meaning from the actual words they say. More than 90% of the meaning is communicated by their tone of voice and through visual clues like their facial expressions, body language and the clothes they are wearing.[1] Your clothes, like your possessions, transmit clues about your personality, financial status, social group, background etc. They tell the world who you are, or who you want to be today.[2] You make judgements about others within the first seconds of meeting based purely on appearances.

'*Didot* is like a person who dresses up to look classy and knows just where to stop with the make-up.' Gwenäelle Barillon

Fonts convey non-verbal information in a similar way to clothes. Your choice of font tells the world how serious you are, and it clues the reader in to your emotions or your intentions before they have started to read. Just like noticing when somebody is dressed inappropriately or their words do not match their body language — if you use a mismatching font you may find your credibility is called into question (see page 60).

Graphology

Handwriting analysis has been considered a science for many years. We each have our own individual style which, according to the British Institute of Graphology, reveals the 'pattern of our psychology expressed in symbols'.[3] Graphology has been used as an evaluation tool in a range of situations, including screening potential employees, assessing the suitability of a marriage partner, and forensic graphology, which is the study of ransom notes and blackmail demands. Today a great deal of what you write is done by tapping on a keyboard where the fonts you select, and how you use them, replace your personal handwriting style. Your choice of fonts may not be as individual to you as your handwriting but, as already shown, it still reveals a great deal about you — and this can be analysed.

Use fonts and influence people

If you want to really connect with your audience then it is important to consider who they are when you choose a font. This is like speaking in the right tone of voice; for example, you would not talk to your boss in the same way that you would talk to young children. In the business world, using type styles appropriate to the sector can help you to speak the same language, although this is not to suggest that you actually go as far as imitating the client's corporate font. If you are not sure what these are then look up some companies in the sector and see what type styles they use.

To continue the parallel with clothes, fonts can act like an interview suit if you are writing a CV or a letter to a potential client and need to be appropriate both for the industry and the occasion. This is not the time to parade your individuality with an outlandish outfit or a 'personality' font. Remember how Phil Renauld's grades improved when he used a more suitable font (see page 59)?

Influence yourself

Fonts influence what you read: you know whether what you are about to read will be intellectual, informative, silly or important. Fonts set the tone and create a mood that can range from calm to energising. They prime you to know whether what you are about to read will be entertaining or educational, and just how believable it might be.

Choosing fonts to influence yourself: how do each of these examples make you feel?

(a) *Futura*

Temperatures and the general feel of the weather will be a talking point early in the week because it will be windy and cool with plenty of showers but also some drier, brighter spells. The brisk winds will gradually ease as the low pressure centre that's

(b) *Times New Roman*

Temperatures and the general feel of the weather will be a talking point early in the week because it will be windy and cool with plenty of showers but also some drier, brighter spells. The brisk winds will gradually ease as the low pressure centre that's driving our weather eases away

(c) *Century Gothic*

Temperatures and the general feel of the weather will be a talking point early in the week because it will be windy and cool with plenty of showers but also some drier, brighter spells. The brisk winds will gradually ease as

(d) *Bookman Old Style* bold

Temperatures and the general feel of the weather will be a talking point early in the week because it will be windy and cool with plenty of showers but also some drier, brighter spells. The brisk winds will gradually

Could you use this to your own advantage? Could you use fonts to influence yourself?

Copywriter Michael Everett changes his typeface depending on the tone of voice he wishes to write in because he finds it helps him to use the right language style for a project.

For short and snappy car advert slogans he writes in the modern, geometric *Futura*. He uses *Times New Roman* for long and informative editorial articles when he wants to write in a more intellectual style. He creates his invoices in the open and 'easy to read' *Century Gothic*; I wondered whether he was instinctively choosing an 'easy-to-do' font to make his invoicing feel like less of a chore (see page 61).

Try this out for yourself. Open a document or an email and see how different fonts change the tone of what you're writing. A typeface like *Century Gothic* could help you to explain something difficult with clarity. *Garamond* or *Book Antiqua* could help you to feel like you're tapping into your knowledge and wisdom, whilst a solid style like *Bookman Old Style* bold might help your words to flow more assertively as you write.

But remember you have chosen the font for yourself. Before you send out the document think about the recipient, and whether this is the right style for them to read your message in. Some fonts may be best used in the privacy of your own home.

Font Fortunes

At the beginning of the book you were invited to make your choice from the selection of Font Fortunes. This is something that began as a fun game at a summer event as a way of engaging people in conversation. I wanted to explore the idea that we are drawn to typefaces that reflect our values and aesthetics and the best way to start was by asking questions. I created a selection of goody bags with different fonts on the labels and visitors were invited to choose the one they felt most drawn to. Inside they found a bag of alphabet sweets and a tongue-in-cheek personality analysis.

The analysis is written purely to reflect the associations and provenance of the chosen font, but it has been surprising how closely people thought they reflected their own personalities. It has been interesting to talk to people about their

font choices, and to hear what they think is so accurate about their 'fortune'. These have proved popular and they now make a regular appearance at Type Tasting events.

Type Dating Game

'Is it fair that typography is seen as the least sexy design discipline?' Rob Alderson *It's Nice That*[4]

To expand on the Font Fortunes idea I created the first version of an online Type Dating Game. This is a light-hearted, gamified survey created to encourage participants to think about the personalities of different fonts and which ones they identify with.

Players are invited to choose a typeface to represent themselves at a speed dating event. They then go through three stages of choosing which fonts they would date, ditch and be just friends with. Along the way they are asked why they are making some of their choices. Their explanations, although anthropomorphised, combine to create detailed profiles of the different font personalities.

The names of the typefaces are only revealed at the end so the answers given throughout the game are based on the letterforms. Once they have chosen their final date participants receive personality analyses both of themselves and their date. Like the Font Fortunes game, these are also based entirely on the attributes of the typefaces. When asked how accurate they thought the analyses were, 48% answered 'very' to 'spot on', 38% said 'good' to 'fairly', with just 14% saying they were not accurate.

This is a fun way of exploring whether we pick typefaces that reflect our values and mirror our personalities. It is not particularly scientific, but it does demonstrate that we readily identify with fonts and that we instinctively know which ones we might be compatible with.

Turn the page if you would like to play the Type Dating Game. Which typeface will you be, and which will you date, ditch or just be friends with?

A selection of the results can be found on pages 96–99. At the time of writing 5,145 participants have completed the survey. The majority being non-designers from the UK and US.

Type Dating Game: which font is your dating type?

You arrive at a speed dating event. Decide which name badge best reflects you; choose carefully as this will form everybody's first impression of you.

'Hello' **A**	'Hello' B	**'HELLO'** **C**	Which badge do you choose? ☐ Why? - - - - - - - - - - - - - - - - - - -
'Hello' ***D***	'Hello' E	*'Hello'* *F*	- - - - - - - - - - - - - - - - - - - - - - - - - - - - - - - - - - - - - - - - - - - - - - - - - - - - - - - - -
'Hello' G	**'Hello'** **H**	'Hello' I	- - - - - - - - - - - - - - - - - - - - - - - - - - - - - - - - - - - - - -

First round

You meet your first three potential dates and chat to each for 60 seconds.
Think about their personalities. What qualities are you attracted to?
Who do you fancy?
The bell goes. It's time to fill in your scores to the right ...

DATE #1   **Date #2**   Date #3

Date: ☐
Ditch: ☐
Friend: ☐

Second round

You meet your next three potential dates.
The bell goes. It's time to fill in your scores to the right ...

Date #4   *Date #5*   Date #6

Date: ☐
Ditch: ☐
Friend: ☐

Third round

You meet your final three potential dates.
The bell goes. It's time to fill in your scores to the right ...

Date #7   **Date #8**   Date #9

Date: ☐
Ditch: ☐
Friend: ☐

Time to decide. Tick who you choose as your ultimate date?

#1 ☐   #2 ☐   #3 ☐   #4 ☐   #5 ☐   #6 ☐   #7 ☐   #8 ☐   #9 ☐

Why did you choose this person?

- -

- -

Turn to page 99 for the personality analyses.

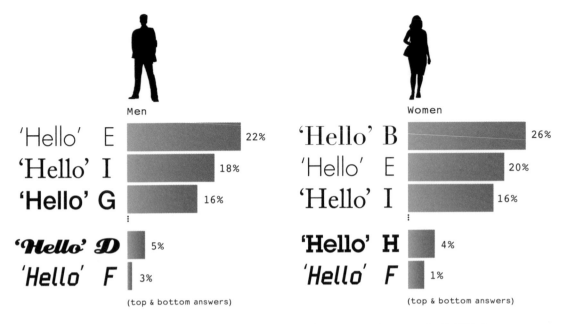

Men

'Hello' E — 22%
'Hello' I — 18%
'Hello' G — 16%
⋮
'Hello' D — 5%
'Hello' F — 3%

(top & bottom answers)

Women

'Hello' B — 26%
'Hello' E — 20%
'Hello' I — 16%
⋮
'Hello' H — 4%
'Hello' F — 1%

(top & bottom answers)

Why men choose to be E (*Futura*).

It's simple and easy-going. Slick. Simple and well-polished. Tasteful, modest, classic. Minimalist, but substantive. Elegant, legible. Practical, clear, elegant. Smart, simple, open, secure and true. Edgy, clean, smart. Serious, old school, simple. Genuine. Down to earth. Normal and standard. It shows power and class without arrogance. Low key. Quiet, unassuming. Shows a bit of interest and personality. A modern classic. Modern, slick, urbane, polished. Reserved and subtle. It's crisp and sharp-looking, kind of like how I like to portray myself. It shows I'm no nonsense …

Why men choose to be I (*Caslon*).

It's kinda serious, but not too serious. Kinda fun, rounded bit at the end of the 'y'. Classic, balanced, meritocratic. Bookish, intellectual, reserved but interesting. Very open and easy-going. Reserved, classical, serious, meticulous. Straightforward, but a touch of flare. Intelligent. Classical elegance. It is straightforward, no-nonsense and classy. Traditional yet bold. Grounded, classy, clear. Classic, clear, simple but thoughtful, invites exploration. Uncomplicated without being plain. Clean, down to earth, no frills. Clean, not overbearing, direct. Formal but not overly so …

Why women choose to be B (*Didot*).

Serious and intellectual yet can be fun. I like beautiful classic things. Subtly classic. Fun & flirty, but stable & loyal. Calm, precise, clean. Sophisticated. Love of beauty, precision and intelligence. Classy, feminine, smart, stable. Sophisticated but sexy. To the point with flair. Classic. I think it's classy and elegant. Subtle self-confidence. A little personal flare. Classic, not flashy. It's very measured. Honest, sense of humour. Elegant style. Very simple, fairly reserved. Easy-going, laid-back. Intelligent. It's elegant and differentiated. Classic, polished and elegant …

Why women choose to be E (*Futura*).

Neat and simple. It's round and organised and looks neat. Clear and straightforward. Sleek, crisp and to the point. It is a little unusual, contained and shapely. Simplistic and not too flashy. It's simple but sexy. Not fussy, clean, somewhat modern. Cute, clean, simple. Honest, classic. I am down to earth, but interesting. Easy, not intimidating. Simple, straightforward. Open and unpretentious, not 'in your face'. I want you to make your own impression without me telling you what it should be. Clean and crisp lines. Simple, no drama, minimalist. It's clear simple and classic …

Results: which are the most dateable types?

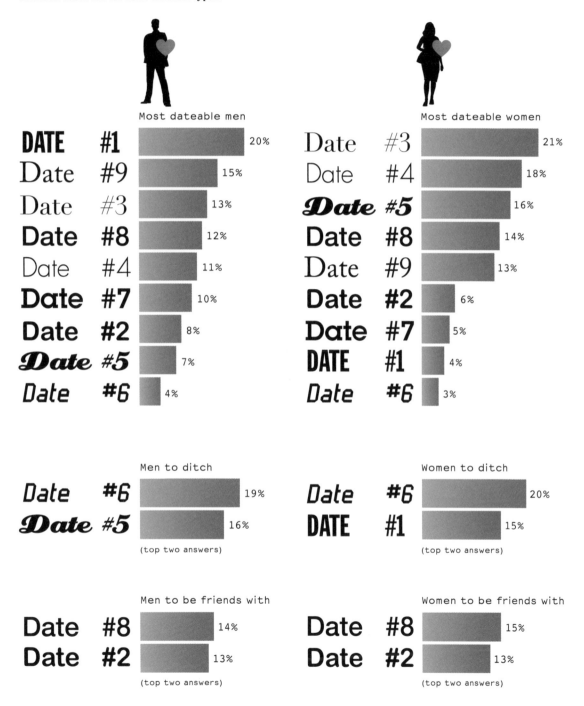

Most dateable men

DATE **#1** 20%
Date #9 15%
Date #3 13%
Date **#8** 12%
Date #4 11%
Date **#7** 10%
Date #2 8%
Date #5 7%
Date *#6* 4%

Most dateable women

Date #3 21%
Date #4 18%
Date #5 16%
Date **#8** 14%
Date #9 13%
Date **#2** 6%
Date **#7** 5%
DATE **#1** 4%
Date *#6* 3%

Men to ditch

Date *#6* 19%
Date #5 16%

(top two answers)

Women to ditch

Date *#6* 20%
DATE **#1** 15%

(top two answers)

Men to be friends with

Date **#8** 14%
Date **#2** 13%

(top two answers)

Women to be friends with

Date **#8** 15%
Date **#2** 13%

(top two answers)

How the most dateable typefaces are described

Most dateable men

Strong and masculine and well dressed. Friendly & easy-going. Strong straightforward and bold. Good-looking, smart, refined. The font is masculine and confident. Tall, bold, and can be fun. Not obtrusive. Looks romantic. A little out-of-the-box, but not too much. Classic, gentleman, intelligent. Responsible, yet fun. Well balanced, interesting, funny. Easy to get along with, not overbearing and good with almost any situation. He looks good in a suit & tie, or in jeans & a t-shirt. Can be serious & have fun. Attractive, funny, intelligent, decided, adventurous, masculine. Solid, bold. Sincere and normal. Decisive and charismatic. Bold and thorough, a good man as a role model. Approachable, trustworthy. Strong yet not too serious. Looks smart and goal-oriented. Hard-working, strong, sexy, manly, great father, great friend, great lover. He seems like a good guy. Stands out, bold, strong. Attractive, well-rounded, interesting. Strong but artsy. Masculine, easy to understand, classic, steady. Smart, with money, nice. Honest and sincere. He's bold, smart, confident. Confident. Kind. Honest. Strong straightforward and bold. Intelligent ...

Most dateable women

She is fun, sexy and casual. She is intelligent and stable. Quiet yet sophisticated. Elegant, attractive. Scholarly, professional, and feminine. Calm and beautiful. She looks fun! Seems trusting, has good feminine qualities, takes care of herself. She seems both trustworthy and interesting. Educated and sophisticated. Beautiful, cultured. Classical, a little flirty lots of stability and traditional. Classy, likes to dress up and go out nice. Classic elegance with style and substance. Beautiful, intelligent, slightly reserved. Sophisticated but sexy. A little untraditional, unshowy, simple. Probably best ballroom dance partner. She seems a bit quirky. Stable, classy, timeless, understated. Easy-going, but also gorgeous. Exotic, sensual, fun and a little dangerous. Clean, feminine, and might be fun to figure out a way to take out of her comfort zone. Sexy. She has a nice body, nice curves. The skinniest one. Friendly, sophisticated, casual. Clean, classy, and modern. Stylish, but not begging for attention. Classic, beautiful and multifaceted. Classy. Curved, definite, bold. Kind of old-fashioned, but still fun. Elegant, classy, feminine. She could be classy ...

How accurate do people say their personality analysis is?

Very. Pretty correct. Pretty spot on. Somewhat. Accurate except I'm not that fancy. Not very accurate. VERY accurate!! 70%. scary :) Very. Pretty cool, but I love to entertain people tho. Very! Not bad. Somewhat accurate. Not Even Close. Very. Somewhat. My analysis is right on! Cool and scary lol. Spot on. My date is exactly what I look for in a partner and I would like to think I'm a very relaxed and laid-back person. Dead on. Accurate. Amazingly accurate on the potential date. This describes my husband to a 't'. Wow! Not very good at articulating at times, but otherwise accurate. Semi-accurate. VERY. 100% spot on for me! A little too spot on maybe! I am a Designer though I know Type! It is mostly accurate for me but exact for my husband. Good. Scarily accurate. Not bad. Not sure. I think it was pretty good, maybe 90% accurate, fun survey! Maybe 70%. I really like it. F***ing accurate! Pretty accurate. I don't know how traditional I am. Perhaps with design, but not so much real life. ACCURATE!!!!! Quite ok :) Fairly. My date is accurate, but my font isn't. 100%. Pretty accurate. Identical ...

Personality analysis given out for each typeface

'Hello' A Date #2

You are friendly and open, usually to be found with a welcoming smile on your face (when you get old you'll have smile lines not frown lines). You're less of an intellectual and more of an appreciator of all things sensory — from the taste of food to creating a comfortable environment. If friends need a comforting voice and a shoulder to cry on they usually turn to you. You chose **Arial Rounded bold** by Robin Nicholas & Patricia Saunders.

'Hello' B Date #3

You relish the opportunity to dress up in style and go to a glamorous party. You enjoy fashion and are drawn to classic and European styles. You appear to be confident and outgoing but this hides your more reserved and possibly introverted personality. You think it's the small, finishing touches that reveal the most about a person. You chose **Didot** by Firmin Didot.

'HELLO' C DATE #1

You have an alpha personality: you're confident and easily take the lead in any situation. You're comfortable taking centre stage but this will be for practical reasons; you're not here to entertain. What people don't always see is that you're actually a big softie with a generous heart. Your family and friends are important to you and you're always there to protect them. You chose **Franklin Gothic bold condensed** by Morris Fuller Benton.

'Hello' D Date #5

You are casual and laid-back, far more comfortable in jeans, t-shirt and bare feet than suited and booted. You are drawn to slightly unconventional people who are often artists or creatives. You're not an organised clock-watcher, preferring that events unfold in their own time. You chose **Eclat** by Doyald Young.

'Hello' E Date #4

You are stylish in your appearance, with a designer aesthetic. You speak openly and clearly, preferring to think things through and give a well-considered opinion rather than to just blurt out the first thing that comes to mind. Your home reflects your minimalist chic style. You chose **Futura light** by Paul Renner.

'Hello' F Date #6

You are confident and assertive. You think quickly on your feet and are a dynamic go-getter in all things. You're most likely to by found in active settings both in your work life and your leisure time. You get bored quickly. You like to win. You chose **Cinema Gothic** by FontGeek.

'Hello' G Date #8

You like to appear neutral and to weigh up a situation before revealing your true personality. In social situations you're more of a chameleon than a peacock, preferring to fit in rather than to take centre stage. You chose **Helvetica** by Max Miedinger, Eduard Hoffmann.

'Hello' H Date #7

You are confident and have your feet firmly planted on the ground. You're more of a doer than a thinker. Your opinions are well considered and you balance out the pros and cons, but at the end of the day you call a spade a spade. You chose **Lubalin Graph** by Herb Lubalin.

'Hello' I Date #9

You have traditional values. Although first impressions may be that you are fairly conservative, people quickly discover that you are curious and inquisitive. You are well read and can always recommend a good book or two. You are also well travelled and have many stories to tell. You chose **Caslon** by William Caslon.

References

1 'Silent Messages: A Wealth of Information About Nonverbal Communication (Body Language)' by Albert Mehrabian, Professor Emeritus of Psychology at UCLA, 2009, kaaj.com.
2 'You Are What You Wear: What Your Clothes Reveal About You' by Dr Jennifer Baumgartner, 2012, Da Capo Lifelong.
3 British Institute of Graphology, britishgraphology.org. Mehrabian poses the theory that when we talk 7% of what we communicate is through our words, 38% by our tone of voice and 55% by our body language.
4 It's Nice That, itsnicethat.com.

Online Type Tasting surveys can be found at typetasting.com.

Sniff this
7
Sniff this
8
Sniff this
9
Sniff this
9
Sniff this
8

11

Sensory type

Props from a Type Tasting session

SENSORY TYPE

Having the ability to absorb information through multiple senses simultaneously speeds up your ability to judge situations and to react quickly when necessary. Neurobiologists Stein, Stanford and Rowland suggest that this is fundamental to your ability to recognise signals and to communicate, and that this has played a vital role in human survival.[1]

Today you live in a visually dominant world and almost half of your brain is involved in processing what you see, yet biologically you are built to use all of your senses to interpret your environment. For example, a large proportion of your genes is devoted to detecting odours, which suggests that smell played a much more important role in your evolutionary past.[2]

Training expert Dugan Laird breaks down how we learn through our senses: 75% through seeing; around 13% from hearing; the other 12% through touch, smell and taste. However, if more than one sense is stimulated at any one time, the experience can become intensified and the learning is considerably more effective. Engaging multiple

senses can be used as a powerful teaching tool and memory aid.[3]

Your strongest memories are multisensory, and this can be demonstrated by asking you to describe (out loud) a past holiday experience that sticks in your mind. You will most likely refer to more than one of your senses: what you saw, the temperature, what you touched and how it felt, what you could smell and often what you ate or drank.

Reading a printed book is a multisensory experience: you smell the paper, feel the texture and hear the pages as you turn them. After reading her first book on an eReader a friend told me that, although it was a good book, it had lacked impact and she felt like 'the intensity had been turned down'. She realised she had missed tactile qualities like the feel and smell of the paper.

Tasting the rainbow

'Why does the voice of the ventriloquist appear to come from the lips of his/her dummy?' asks Professor Charles Spence.[4] He explains that not only does perception occur simultaneously across all of your senses, but what you experience in one sense can trigger a reaction in another ('sensory modality'). These are known as 'crossmodal' interactions. For example, you see a colour and know what it would taste like; you hear water pouring and know what temperature it would be; a smell triggers the anticipation of a flavour; you know what a shape would feel like to touch; or how loud and shrill a font would sound. The quiz opposite demonstrates how automatically you do this.

Sarah Hale tells of starting a new job working in a supermarket staff canteen. An 'old-hand' cook explained the recipes to her, including one for raspberry sauce made with sugar, water, cornflower and pink colouring. 'But where's the raspberry?' Sarah asked, and says 'A look of amazement stole over the cook's face as if this had never occurred to her. 'I dunno.' she said, 'it *tastes* of raspberry.''[5]

Some of your instinctive responses to shapes, taste and colour are hardwired into your evolution for your protection and survival, according to beer expert Pete Brown.[6] For example, fruit turns red and round when it is ripe to eat, so you associate red and round shapes with sweetness. Many poisons are sour or bitter, so your sense of taste has evolved to help you avoid these.[7]

taste sound sight *smell* **touch**

You also learn associations from seeing type used on food packaging and in advertising. You know what indulgently decadent (or low calorie) fonts look like, which are cheap or expensive, and which are most likely sweet or savoury.

Exploring the crossmodal potential for type, or 'sensory type', is something I am involved in researching on an ongoing basis. Some of the initial ideas and outcomes are shown in this book.

Synaesthesia

Approximately 1 in 2000 people have synaesthesia[8] and experience senses that would usually be separate, as linked together.[9] For example, a synaesthete might hear colours, see sounds as shapes; the number five might always be red, or taste metallic. These are crossmodal experiences created by the brain that are unique to that person. Famous artists with synaesthesia include David Hockney and Vincent Van Gogh, Olivier Messiaen, Tori Amos and Duke Ellington.[10]

Lisa Olliff, a 49-year-old synaesthete, only realised five years ago that the way she sees things is unique to her. She told me 'I can't understand how others see things in their heads.' When Lisa looks at months, days and letters, they have colours. For example 'July is bright green, October is white, H is yellow, Wednesday is burnt orange and K is green.' She also sees the calendar in 3D and from different angles: 'It's as if I am viewing from the side or as if I am on a train track, riding on it.'

Graphic designer Jamie Clarke experiences synaesthesia as individual letters and numbers each having a 'correct colour', and these are consistently the same. He says that this does not impede his graphic design work as the effect is 'blotted out' once the characters are combined to create words or sentences.

Crossmodal research

At the Crossmodal Research Laboratory at Oxford University[11], Professor Charles Spence and his team research the links between words, shapes and flavours. They have proved that soft and round shapes are associated with sweet tastes and rounded sounds, like the nonsense word 'maluma'. Jagged shapes taste sour and associate with harsh sounds like the word 'takete'.[12]

'takete'
SOUR

'maluma'
SWEET

They have found that these associations also apply to typefaces; the team have proved that an angular typeface (53-point *Hollywood Hills* regular) scores as tasting sour, and that a rounded typeface (44-point *Swis721 BlkRnd* black) is rated as tasting sweet.[13]

Bitter/sweet typography

The team at Heston Blumenthal's Fat Duck restaurant, along with the Crossmodal Research Laboratory[14], have identified specific sounds that are linked to different tastes. These have been used as the basis for soundscape compositions which have been shown to intensify the experience of each taste when food is eaten whilst listening to them.[15]

Reading about this prompted me to wonder whether the same could be done with type; whether letterforms could be matched to each taste and we could ultimately create typefaces for bitter, salty, sour, sweet and umami. These are my initial explorations to provide a starting point.

Process

1 I selected typical food for each taste and ate it whilst listening to the appropriate soundscape. I filled a sheet of paper with abstract marks and shapes that visually represented the experience of each taste (see right).

2 I took the abstract marks and either sourced typefaces with similar visual qualities or, where a typeface did not already exist, I either modified existing fonts or created new letterforms.

3 To test out my results I created an online survey to see whether participants matched the letterforms to the same tastes. I included images of food to prompt more accurate sense memories.

Results

The results of the online survey show most agreement with the 'sweet' typeface, and this bears out the original findings of the Crossmodal Research Laboratory. The other tastes show some, but less significant, agreement. So, the next stage is to develop the shapes further and to test them in person, with real food.

Future research

I am doing collaborative research with the Crossmodal Research Laboratory to explore this further. They test the proposed ideas under more rigorous scientific conditions; the first study is to be published shortly.[16]

Type Tasting online survey Bitter/Sweet

Sweet

Abstract marks   *Candice*

Salty

Abstract marks   New letterforms

Bitter

Abstract marks   Modified *Klute*

Sour

Abstract marks   New letterforms

Results (answers from 82 participants)

sweet	92%	salty	36%	bitter	38%	sour	39%
bitter	3%	sour	36%	sour	31%	bitter	33%
sour	3%	bitter	25%	salty	26%	salty	23%
salty	2%	sweet	3%	sweet	5%	sweet	5%

Still or sparkling?

The team at the Crossmodal Research Laboratory investigated label designs on a range of still and sparkling water bottles.[17] They conducted a study that looked at both shape and colour on the labels and concluded that shape is more important than colour for differentiating between the two types of water. An organic-shaped logo was consistently matched with the still water, whilst the sparkling water was consistently matched with the angular-shaped logo (right). The team observed that these findings are not reflected in current water-bottle label designs.

Organic   Angular

I became interested in exploring whether these findings could be applied to the typefaces used on water bottles, and the implications this could have for graphic designers in terms of the range of graphic language we use.

1 Do rounded and angular typefaces match the experience of drinking still and sparkling water?

2 Is it possible to create a scale of still to sparkling (calm to energising) typefaces?

Process

I created two online surveys in which participants compared typefaces and letterforms, indicating which they thought were more 'still' or 'sparkling'. After cross-referencing both sets of results I was able to create the still to sparkling scale on the right.

Results

Rounded typefaces are consistently matched to still water and angular ones to sparkling water. The most 'still' typefaces have balanced shapes, with no thick/thin contrast. This is more important than whether the letter shapes are curved or angular. The transition from still to sparkling can be broken down into five distinct stages.

Still   Sparkling

Sparkling (energised)

Sloping forwards, not flowing
Thick/thin contrast
Angular shape

water

Sloping forwards, not flowing
Moderate thick/thin contrast
Rounded and angular shape

water

Sloping forwards, flowing
Moderate thick/thin contrast
Rounded and angular shape

water

Balanced
Moderate thick/thin contrast
Rounded, organic shape

water

Balanced
No thick/thin contrast
Rounded, geometric shape

Still (calm)

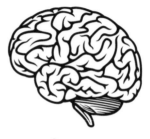

Higher up the price range the typography on the packaging becomes more intellectual.

Cheap and instant food packaging often appeals directly to the desires of your subconscious.

Judging a meal by its packaging

Today we rarely hunt for, or gather, our food in the wild; we purchase most of it which means that it is linked to decisions about which shops we go to and the brands we buy. According to design writer Steven Heller, 'Food is culture, it is a code or symbol for class and social strata' and the packaging is designed to both appeal to, and reflect, your lifestyle. He comments that 'We are what we eat' is a familiar phrase, but today 'We are what we buy' would be more accurate.[18]

In some instances the making of food choices as a lifestyle statement can even overshadow decisions about the nutritional value of the food itself. 'From sophisticated choices to culinary camp, your choice of food proclaims your attitude towards life,' says Catharine Weese.[19] We expect food to both entertain and nourish us and we effectively consume a meal twice: once in our imagination when we're in the shop and again later when we actually eat it.[20]

'The drama of the package will always give you a better appreciation of the product.' Packaging designer Primo Angeli

Gastroporn

Cheap and instant food packaging often appeals directly to the desires of your subconscious. These are the instincts that are hardwired into your evolution. The food packaging entices you with curvaceously sweet-looking letters, words written salaciously in melted cheese, or drizzled out of rich and creamy chocolate. The words are often in the shape of the food they describe, so that the lettering does not just say cheese, it *is* cheese.[21] These designs tap into your primitive survival instincts, which drive you to search for sweet and calorific foods.

Higher up the price range the typography on the packaging becomes more intellectual, and you must refer to your mental database of learned associations. The type is generally more understated, often in traditional Roman Serifs, italics, lightweight typeface styles and sometimes combined with handwriting or distressed effects to signify 'hand-made'. Generally, the more expensive it is, the more refined the type becomes. Packaging conventions vary from country to country, so when you are shopping abroad, where the shops are unfamiliar, you can end up with a few surprises in your shopping basket.

Haddock and chips

A: *Balega*

Haddock
and chips

B: *Helvetica Neue* thin

Haddock and chips

C: *Gill Sans* ultra bold

HADDOCK AND CHIPS

D: *Gill Sans Shadow*

Describe the style and price range of each meal on the left from the type.

A:

B:

C:

D:

Turn to page 108 to compare your answers with the key to food packaging type.

beer

We are used to seeing angular typefaces on beer labels.

Love it or hate it, the London 2012 typeface energised the Olympic Games that year.

We don't like triangles

The brain's fear processor, the amygdala, plays a key role in recognising potential threat.[22] It is activated by visual elements with angular or sharp contours, and remains unaffected by soft contours and round shapes. As a result we favour curve-shaped objects; this is called 'contour bias'.

Packaging designers are aware of contour bias and this is why most packaging features either curved letters or classic Serif/Sans Serif type styles. This is even the case on products like crisps, where angular shapes and lettering might better represent the crunchy experience. Lettering styles are generally rounded and 'friendly' (although they might occasionally feature a triangle in the background of the design or in the logo).

Marketing psychologist Louis Cheskin was the pioneer in studying people's emotional responses to packaging in the 1930s.[23] In his most famous experiment he found that 80% of people preferred a design with circles when given a choice between circles and triangles. Manufacturers are wary of incorporating angular shapes, or making any other changes, that might reduce sales.

In his book about hidden meanings in packaging,[24] Thomas Hine explains that designing packaging is a balance of getting a product noticed, but also having it accepted into people's homes. A design that is angular and attention-grabbing in the shop may not convey the right emotional qualities once it is in the domestic environment. Cheskin demonstrates that triangles are highly visible, although 'people can see them but that doesn't mean they like them'. He found that circles and ovals are received most positively, but that they are found to lack personality on their own, and become more interesting when combined with angular shapes.

However, contour bias only holds true for neutral objects and shapes, it is dispelled once you recognise the object. A hand grenade is a safe, round shape, but you have learned that it is dangerous, which overrides the amygdala's instinctive (lack of) response.

Designers and manufacturers could take more risks that would result in the typographic landscape of the supermarket becoming more interesting. After all, we are used to seeing angular type styles on beer labels and — love it or hate it — the London 2012 typeface energised the Olympic Games that year.

Key to food-packaging type (from Type Tasting online survey results)

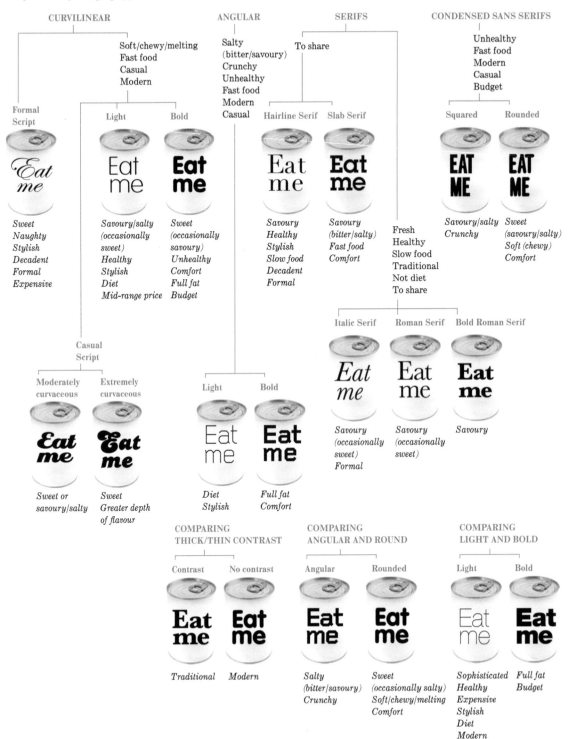

CURVILINEAR

Soft/chewy/melting
Fast food
Casual
Modern

Formal
Script

Sweet
Naughty
Stylish
Decadent
Formal
Expensive

Light

Savoury/salty
(occasionally
sweet)
Healthy
Stylish
Diet
Mid-range price

Bold

Sweet
(occasionally
savoury)
Unhealthy
Comfort
Full fat
Budget

Casual
Script

Moderately
curvaceous

Sweet or
savoury/salty

Extremely
curvaceous

Sweet
Greater depth
of flavour

ANGULAR

Salty
(bitter/savoury)
Crunchy
Unhealthy
Fast food
Modern
Casual

Light

Diet
Stylish

Bold

Full fat
Comfort

SERIFS

To share

Hairline Serif

Savoury
Healthy
Stylish
Slow food
Decadent
Formal

Slab Serif

Savoury
(bitter/salty)
Fast food
Comfort

Fresh
Healthy
Slow food
Traditional
Not diet
To share

Italic Serif

Savoury
(occasionally
sweet)
Formal

Roman Serif

Savoury
(occasionally
sweet)

Bold Roman Serif

Savoury

CONDENSED SANS SERIFS

Unhealthy
Fast food
Modern
Casual
Budget

Squared

Savoury/salty
Crunchy

Rounded

Sweet
(savoury/salty)
Soft (chewy)
Comfort

COMPARING
THICK/THIN CONTRAST

Contrast

Traditional

No contrast

Modern

COMPARING
ANGULAR AND ROUND

Angular

Salty
(bitter/savoury)
Crunchy

Rounded

Sweet
(occasionally salty)
Soft/chewy/melting
Comfort

COMPARING
LIGHT AND BOLD

Light

Sophisticated
Healthy
Expensive
Stylish
Diet
Modern

Bold

Full fat
Budget

1933

Guns

Bold, large condensed e.g. *Alternate Gothic*

Cigars

Bold, non-italic, condensed e.g. *Cheltenham condensed extra bold*

Moving pictures

Regular, large e.g. *Caslon* bold

Typewriters

Non-italic, small e.g. *Century*

Golf clubs

Bold, medium, non-italic e.g. *Cheltenham bold*

Perfume

Non-bold, regular or small, italic e.g. *Goudy Old Style* italic

1935

Cheapness

Cheltenham bold

Dignity

Caslon

Luxury

Savoye

Strength

Cheltenham bold

Economy

Goudy

Automobiles

Caslon italic

Perfume

Savoye

Coffee

Goudy

Jewellery

Savoye

Building materials

Cheltenham bold

Type styles for historic commodities

'A typeface needs to be sympathetic to the product and to emotionally engage with the consumer.' Michael Everett, copywriter

Some of the early research into typefaces was conducted in 1933 by psychologists R. C. Davis and Hansel J. Smith at Indiana University. In their study, called the 'Determinants of feeling tone in type faces'[25], the psychologists explain that they had observed statements like 'bold type expresses cheapness', 'italic types express femininity', and '*Bodoni* type expresses modernness', being made regularly, but without proof. They set out to prove the validity of these claims, and to match fonts to a range of 1930s commodities, from guns through typewriters and cigars to perfume, as a guide for graphic designers and advertisers.

Two years later Gwendolyn Schillebi opened her report in the *Journal of Applied Psychology* with the statement: 'Advertising is one of the most important fields to which psychology has been applied.' In her study, 'An experimental study of the appropriateness of colour and type in advertising', participants were asked to match typefaces to abstract qualities like cheapness and dignity, and then to commodities like automobiles, coffee and building materials.[26]

A selection of the results of both studies is illustrated above (typefaces have been substituted where the originals were not available).

Packaging design today is more sophisticated than it was a century ago, in terms of both materials and design. However, taking into account that Sans Serif type styles were not yet in common use in the 1930s, these examples show that typographic conventions have changed little in over 80 years.

Wake up and smell the fonts

We interact with fonts either as professionals who actively use typefaces, or as type consumers on the receiving end.

Type professionals
A type professional is engaged in the process of consciously choosing and using type in their work and, as a result, they have a wide vocabulary to draw on. Non-professionals can find this language intimidating and, when I start conversations, people will often begin with an apology, 'I'm sorry I don't know anything about fonts.' There is a prevailing misconception that people need to be experts to have opinions about type.

Type consumers
As the end users of the design process we encounter fonts as we go about our everyday lives. We generally 'look past the font' and interact with them unconsciously. We are all experts as type consumers; we all know instinctively how to respond to the different shapes and styles of fonts because we have been doing this all our lives. Fonts only communicate with us because we already understand their meanings.

In the film short *Font Men*, type designer Jonathan Hoefler explains that he thinks there is a 'poverty of descriptive terms' for the experience of typography. He finds he uses phrases that are purely descriptive, or that involve cultural references, like: 'It's too Steven Segal and not Steve McQueen enough'.[27]

Part of what I do under the Type Tasting banner is initiate conversations about the experience of typography from the type consumer's point of view. This involves getting away from the jargon and using the descriptive language of the consumer, which can range from discussing which fonts you would date, to how they might taste or smell.

In a Type Tasting session participants are set challenges which include being asked to match jars of smells and bowls of food stuffs to a range of typeface samples, or to play games involving typeface personalities. The debates that follow can be very lively and, although they are about taste, texture and smell, the language used sounds authentically typographic.

Answers
Page 102

Typefaces
Page 103

Five senses typefaces: *Candice, Shatter, Futura, Edwardian Script, Klute.*

Page 105
Still: *VAG Rounded*, Sparkling: *Klute*.

Sparkling to still: *Klute, Cocon (Mateo, Cocon, Bodoni Poster* italic), *Cocon (Cocon, Bodoni Poster* italic, *Candice), Candice (Cocon, Bodoni Poster* italic, *Candice), VAG Rounded.*

Page 107
Beer: *Fette Fraktur*, Olympics: *London 2012.*

References
1 'The Neural Basis of Multisensory Integration in the Midbrain: Its Organization and Maturation' by Barry E. Stein, Terrence R. Stanford, and Benjamin A. Rowland, 2009.
2 *The ICI Report on the Secrets of the Senses*, 2002, The Communication Group, London.
3 *Approaches to training and development* by Dugan Laird, 1985, Perseus Books.
4 'Crossmodal processing' by Charles Spence, Daniel Senkowski and Brigitte Röder, 2009, Crossmodal Research Laboratory.
5 'Flavour: there's a lot more to it than meets the eye. (Or tongue, or nose, or ears...)' by Pete Brown, 2013, petebrown.blogspot.co.uk.
6 Ibid. Pete Brown.
7 'A Matter of Taste', by Society for Neuroscience, 2011, brainfacts.org.
8 'Everyday fantasia: The world of synesthesia' by Siri Carpenter, 2001, apa.org.
9 UK Synaesthesia Association, uksynaesthesia.com.
10 'List of people with synesthesia', en.wikipedia.org.
11 Crossmodal Research Laboratory, Department of Experimental Psychology, University of Oxford.
12 'The Sweet Taste of Maluma: Crossmodal Associations Between Tastes and Words' by Anne-Sylvie Crisnel, Sophie Jones, Charles Spence, 2012, Crossmodal Research Laboratory.
13 'Predictive packaging design: Tasting shapes, typefaces, names, and sounds' by Carlos Velasco, Alejandro Salgado-Montejo, Fernandon Marmolejo-Ramos, Charles Spence, 2014, Crossmodal Research Laboratory.
14 'As bitter as a trombone: Synesthetic correspondences in nonsynesthetes between tastes/flavors and musical notes' by Anne-Sylvie Crisnel, Charles Spence, 2010, Crossmodal Research Laboratory.
15 'A bittersweet symphony: Systematically modulating the taste of food by changing the sonic properties of the soundtrack playing in the background' by Crisinel A-S, Stefan Cosser, Scott King, Russ Jones, James Petrie, Charles Spence, 2012, *Food Quality and Preference.*
16 'The taste of typeface design' by Carlos Velasco, Andy Woods, Sarah Hyndman, & Charles Spence, 2015, Crossmodal Research Laboratory.
17 'On the colour and shape of still and sparkling water: Insights from online and laboratory-based testing' by Mary Kim Ngo, Betina Piqueras-Fiszman, Charles Spence, 2012, Crossmodal Research Laboratory.
18 'Food Fight' by Steven Heller, 1999, *AIGA* vol. 17.
19 *Design as a Main Course* by C. Weese, 1999, *AIGA* vol. 17.
20 'Eat your words: Food as a system of communication' by Sarah Hyndman, 2001, London College of Communication.
21 Ibid. Sarah Hyndman.
22 'Visual elements of subjective preference modulate amygdala activation' by Moshe Bar and Maital Neta, 2014, *Neuropsychologia.*
23 *The Total Package: The Secret History and Hidden Meanings of Boxes, Bottles, Cans, and Other Persuasive Containers* by Thomas Hine.
24 Ibid. Thomas Hine.
25 'Determinants of feeling tone in type faces' by R C Davis, Hansel J Smith, 1933, Indiana University.
26 'An experimental study of the appropriateness of colour and type in advertising' by Gwendolyn Schillebi, 1935, Barnard College.
27 'Font Men' by Dress Code presented by *AIGA*, 2014, vimeo.com.

Online Type Tasting surveys can be found at typetasting.com.

low calorie

Cooper Black / Helvetica light

STRONG
POWERFUL

Flemish Script / Helvetica condensed black oblique

VAG Rounded bold / *Rubber Stamp*

Candice / Modified *Klute*

water

Klute / *VAG Rounded* bold

12

Hold the book up to a mirror to reveal what the blue words are spelling out.

Altered experiences

Mateo / *Bodoni Poster* italic

Monotype Script / *Bauhaus* light   *Klute* / *Aspirin*   *Candice*   *Drawn letters*

VAG Rounded

Experiment: Sweet and Sour

You will need two matching fruit-flavoured sweets—for example, jellybeans or fruit gums of the same flavour and colour.

Eat Me #1

Eat the first sweet while you look at the words 'eat me', above. Score how it tastes. Now turn to page 116.

How sweet? (0 = not sweet, 10 = extremely sweet)

0	1	2	3	4	5	6	7	8	9	10
☐	☐	☐	☐	☐	☐	☐	☐	☐	☐	☐

How sour? (0 = not sour, 10 = extremely sour)

0	1	2	3	4	5	6	7	8	9	10
☐	☐	☐	☐	☐	☐	☐	☐	☐	☐	☐

Before reading this chapter, take part in the Sweet and Sour experiment on pages 114 and 116.

'It's not the mouth but the brain that governs what foods and drinks we like or dislike' explains chef Heston Blumenthal. We absorb a great deal of information about what we eat through our different senses but it is the brain that makes sense of this information. What's happening on your tongue is affected by what's happening in your head, and this means that your tastebuds are 'susceptible to influences from many sources'.[1] In other words, your expectations can alter the actual experience of the food you eat.

In his TEDx talk, Professor Charles Spence explains that if you tell somebody that a bottle of wine is either £90 or £5, they will experience the £90 bottle as being more pleasurable to drink, even if it is the same bottle of wine. He says they will also find the aspirin they take the next morning

to be 'significantly more effective' in reducing the symptoms of pain if they are told the packet cost £5 instead of £1.[2]

'The drama of the package will always give you a better appreciation of the product.' Primo Angeli[3]

The first time I experienced this altering of perception was at a demonstration based on Blumenthal and Spence's work.[4] The sound of a crackling fire was playing in the background and an image of pale tree trunks was projected across the wall as we drank a glass of whisky. It tasted deep and smoky. Then the sound of the crackling fire was changed to high, tinkling sounds, and the trees were replaced by a deep red light. Suddenly, mid-mouthful, the whisky became sweet — all the smokiness had gone. They explained that they could alter a person's perception of what they tasted by up to 20% with colour, light and sound.

Dr David Lewis asked two groups to each read from one of two identically worded menus describing 'rich and creamy' tomato soup. The only difference was that one was written in *Courier* and the other in *Lucida Calligraphy*. After eating the soup they were asked to rate it for taste, freshness and enjoyment. Lewis found that those who read the *Lucida Calligraphy* menu rated the soup as tasting 64% tastier, fresher and more enjoyable. Twice as many said were likely to purchase it themselves. The typefaces primed participants to have a better (or worse) experience of the soup.[5]

Rich &
Creamy
Tomato
Soup

Courier

Rich &
Creamy
Tomato
Soup

Lucida Calligraphy

The emperor's new packaging

In his work dating back to the 1930s, marketing psychologist Louis Cheskin proved that the appearance of the packaging has a dramatic

Klute

Eat Me #2

Eat the second sweet while you look at the words 'eat me', above. Score how it tastes.

How sweet? (0 = not sweet, 10 = extremely sweet)

0	1	2	3	4	5	6	7	8	9	10
☐	☐	☐	☐	☐	☐	☐	☐	☐	☐	☐

How sour? (0 = not sour, 10 = extremely sour)

0	1	2	3	4	5	6	7	8	9	10
☐	☐	☐	☐	☐	☐	☐	☐	☐	☐	☐

Ice Cream A

Ice Cream B

Ice Cream C

Ice Cream D

Ice Cream E

Ice Cream F

Ice Cream G

ICE CREAM H

Which is the sweetest? ____ Which is sorbet? ____ Most expensive? ____ Cheapest? ____

Low calorie? ____ Most rich and creamy? ____ Most appealing? ____

effect on the taste of crackers, soup and beer. He calls the transferring of the imagined experience promised by the packaging to the actual experience of eating the food 'sensation transference'.[6]

How does packaging influence us without us being aware of it? Author Thomas Hine suggests 'We are all so accustomed to looking past the package that we don't see it even when it is put right in front of our eyes'. He explains that we experience a visual overload in a supermarket, which we cope with by not consciously looking at everything, but we 'still see plenty' because we take it all in subconsciously.[7] According to packaging designer Primo Angeli, packaging contains, protects, informs and advertises the food inside. It must stimulate a craving or hunger by 'seducing our most primal urges'. As food became increasingly mass produced from the 1800s, manufacturers saw the opportunity to use the packaging to encourage people to purchase their products. Today there is a huge amount invested in this specialist industry.

Tested on humans

What were your answers to the ice cream quiz at the top of the page? Generally sweet, rich and creamy foods are associated with curvaceous typefaces (A, D, F and G). Angular shapes suggest sour flavours, like citrus sorbet (B). Lightweight and ornate styles are considered to be more expensive E, G), whilst bolder and mechanical-looking letter shapes appear cheap (D, H). The most appealing ice cream depends on your preferences. You may prefer a full fat, brightly coloured indulgence, or a palette cleansing sorbet.

In a supermarket the 'low calorie' or 'healthy eating' options will often be represented by lightweight typefaces like example E (see page 108). The skinny letters and all the space around them represent the reduced calorie content at a glance. However, when I buy them I already know that I will likely still feel hungry at the end of the meal. I buy them because I ought to, not because I have been seduced into wanting to. Since the packaging sets up expectations that influence the eating experience, I would like the healthy option to also convince me that it will also be satisfying to eat.

The Sweet and Sour experiment on pages 114 and 116 is based on a Type Tasting demonstration I give. At an event I gave the 100 members of the audience two jellybeans to eat, both the same colour and flavour. One was eaten whilst looking at round-shaped letters (page 114) and the other whilst looking at a angular-shaped typeface (page 116). Both were in black on white, so it was just the typeface that was changed and the results were not affected by colour. The audience rated the sweets eaten to the jagged-shaped letters as tasting 11% sourer, and the identical jellybeans eaten to the round typeface as tasting 17% sweeter. The type altered how the sweets tasted.[8]

The results are similar every time I do this experiment. How do your answers compare?

I have also done this experiment with crackers, which tasted saltier and crunchier with varied type. These demonstrations are designed to be theatrical rather than scientific. However, this area is being researched by the team at the Crossmodal Research Laboratory under scientific conditions and their findings bear out the results. Looking at different shapes, whether these are logos or typefaces, can alter our perception of taste. This effect can be amplified once colour, shape, texture and photography are included.

Placebo fonts

We live in a time when obesity, diabetes and heart disease are reaching catastrophic levels and society is becoming increasingly post-culinary as we cook less and consume more pre-packaged food.[9] According to the UK government 'in England, most people are overweight or obese, including 61.9% of adults and 28% of children aged between 2 and 15'[10]. The UK government has guidelines on reducing obesity and improving diet that actively encourage businesses to reduce ingredients (for example salt and fat) that can be harmful if people eat too much of them.[11] 'The Consensus Action on Salt and Health in the UK' proposes that the food industry reduces the salt content of all foods by 50–60% over the next five years.[12]

It has been proved that we have the power to influence the taste of food because the packaging sets up expectations that can alter how the food tastes. Could this be used for good? Could we add sugar, salt and fat to food via the packaging, like a placebo, and reduce the amounts in the actual food, but do this so that our perception of what the food actually tastes like is unaltered?

Research is currently underway into the use of multisensory illusions to enable levels of sugar, salt or fat in food to be reduced without altering the perception of the taste.[13]

Typefaces
Page 117
Ice cream A: *Cooper Black*, B: *Cinema Gothic*, C: *Balega*, D: *VAG Rounded*, E: *Helvetica* light, F: *Candice*, G: *Flemish Script*, H: *Impact*.

References
1 'Pleasure, the Brain and Food', *The Fat Duck Cookbook* by Heston Blumenthal, 2009, Bloomsbury.
2 'Expectations of pleasure and pain' by Professor Charles Spence, 2014, TEDxUHasselt.
3 'Upscale, Downscale: We Are What We Eat' by Ellen Shapiro, 1999, AIGA vol. 17.
4 'Assessing the influence of the multisensory environment on the whisky drinking experience' by Carlos Velasco, Russell Jones, Scott King, Charles Spence, 2013, *Flavour Journal*.
5 *The Brain Sell, When Science Meets Shopping* by Dr David Lewis, 2013, Nicholas Brealey Publishing.
6 *The Total Package, The Secret History and Hidden Meanings of Boxes, Bottles, Cans, and Other Persuasive Containers* by Thomas Hine, 1998, Little, Brown and Company.
7 Ibid. Thomas Hine.
8 Typetasting with Sarah Hyndman and Design Week, London Design Festival at the V&A, 2014, londondesignfestival.com.
9 'Eat your words: Food as a system of communication' by Sarah Hyndman, 2001, London College of Communication.
10 'Policy: Reducing obesity and improving diet', 2013, gov.uk.
11 Ibid. gov.uk.
12 Committee on Medical Aspects of Food, Recommendations made about salt, 1994, actiononsalt.org.uk.
13 *The ICI Report on the Secrets of the Senses*, 2002, The Communication Group, London.

Online Type Tasting surveys can be found at typetasting.com.

IMAGINED

Nutrition Facts

Placebo

Total Sugar 28g	9%
Actual Sugar 23g	
Imagined Sugar 5g	

Nutrition Facts

Placebo

Total Salt 0.53g	9%
Actual Salt 0.43g	
Imagined Salt 0.1g	

ADDITIVES

Imaginary **sweet**

Imaginary **salty**

Imaginary **crunchy**

13

Edible type

Helvetica water biscuits with rosemary

EDIBLE TYPE

I have found turning type into food to be an effective way of initiating conversations about the experience of typography. They often provoke lively debate using evocative and descriptive language. A discussion about, for example, what *Times New Roman* would taste like, or what kind of cheese *Helvetica* would be, still sounds authentically typographic.

Some recipes draw on the provenance of the typeface and involve researching its origins. This is a great way to teach and learn about type history. Alternatively, they take the type consumer as the starting point and represent the experience of interacting with typefaces.

Tasting selection box

I originally created these boxes to celebrate Type Tasting's first Christmas. Each contained three typefaces interpreted as food, along with a chocolate-box style description sheet.[1]

Impact, Helvetica, *Comic Sans*

The *Eye* magazine team reviewed a pack, judging each typeface on its taste, texture, letterspacing and suitability. A few of the team's comments are included here and the full review can be read on the *Eye* blog.[2]

Impact: dark chocolate laced with chilli
Impact is a Sans Serif display typeface with ultra-thick strokes and compressed letterspacing. It was designed by Geoffrey Lee in 1965 for attention-grabbing headlines. The dark chocolate with chilli version is bold and packs an in-your-face punch.

'Great, *Impact* should be dark with a kick.' Jay Prynne

'The richness of the dark chocolate and the spicy hit of the chilli makes an impression on the tastebuds.' Sarah Snaith

Helvetica: savoury biscuits
A neo-Grotesque Sans Serif typeface designed for neutrality, which has achieved ubiquity. It was developed in 1957 by Max Miedinger with Eduard Hoffmann. It has been cooked up as simple, savoury biscuits with little flavour of their own, which can work with any topping.

'Practical and safe with no surprises.' Janet South

'Would have been nice with cheese.' Jay Prynne

Comic Sans: candy melts with popping candy
Vincent Connare created the casual script that people love or hate for Microsoft in 1994. As the name suggests, it is inspired by comic book lettering. *Comic Sans* has been interpreted as brightly coloured, extremely sweet chocolate-style candy filled with popping candy.

'Sweet with undertones of guilty pleasure.' Simon Esterson

'*Comic Sans* deserves better.' John L. Walters

Helvetica

Recipe: *Helvetica* **water biscuits with rosemary**[3]

Helvetica was designed to be highly readable, but neutral. It was intended to be an invisible carrier of words that added no additional meaning of its own through the shape of its letters.

The edible interpretation of *Helvetica* reflects this in the form of savoury water biscuits that are plain enough that they can be topped with any food without influencing its flavour. They have a sprinkling of salt and a dash of rosemary for a Swiss Alpine touch.

The biscuits are cut out using 3D-printed cookie cutters. Alternatively, you could use a printout of your selected letters as a stencil to cut around.

Serving suggestion: a menu staple which is good to have in the cupboard for everyday snacks and meals. Best served with cheese, cold meat or a tasty dip.

Ingredients

200g plain flour
1 tsp baking powder
50g butter, cut into cubes
2 tsp dried rosemary
Sea salt

Method

Preheat the oven to 175°C/350°F/Gas 4. Prepare baking trays by greasing them and lining with non-stick baking paper.

1 Mix the flour, baking powder, butter, rosemary and a pinch of sea salt, until the butter is mixed in.

2 Add 4 tbsp water and mix until the dough comes together. If it feels dry add more water and process until the dough is soft but not sticky.

3 Roll out the dough on a floured work surface as thinly as possible. The thinner the better.

4 Cut out your biscuits using either the biscuit cutters or templates you prepared earlier.

5 Place on the baking tray. Brush water over the surface of the dough, sprinkle a little salt over the top. Prick the biscuits with a fork.

6 Bake for 10–15 minutes until the biscuits feel dry but are still pale. Place on a wire rack to cool.

Baskerville

Recipe: *Baskerville* Earl Grey tea biscuits[4]

Baskerville is a Transitional Serif typeface created by printer and type designer John Baskerville in the 1750s. It is a recognisably English typeface that has stood the test of time as a legible, everyday text face.

Baskerville was created at a time when improved technology and transport enabled new and exotic foods to be imported, and tea became the 'national drink'. The edible interpretation of *Baskerville* is as Earl Grey tea biscuits, for an authentic, 18th-century flavour.

The biscuits are cut out using 3D-printed cookie cutters. Alternatively, you could use a printout of your selected letters as a stencil to cut around.

Serving suggestion: these are elegant biscuits with depth of flavour, making them versatile enough as a snack or for more formal occasions. Best served with a pot of tea.

Ingredients

200g butter
200g caster sugar
2 tsp finely ground Earl Grey tea leaves (use decaffeinated tea for less of a caffeine hit)
1 large free-range egg, lightly beaten
400g plain flour

Method

Preheat the oven to 175°C/350°F/Gas 4. Prepare baking trays by greasing them and lining with non-stick baking paper.

1 Cream the butter and sugar with the finely ground tea using a wooden spoon. Add the egg and beat until combined. Add the flour and mix until the mixture forms a dough.

2 Gather the dough into a ball, wrap it in cling film and chill in the fridge for an hour.

3 Roll out the dough on a floured work surface to approx 5mm thick.

4 Cut out your biscuits using either the biscuit cutters or templates you prepared earlier.

5 Place on the greased and lined baking tray. Wash with beaten egg and sprinkle with tea.

6 Bake for about 8 minutes until golden brown. Place on a wire rack to cool.

Burlingame

Recipe: *Burlingame* pick 'n' mix sweets[5]

Monotype, together with The Massachusetts Institute of Technology (MIT), conducted an exploratory study into type styles which would minimise a driver's glance time when looking at in-vehicle displays.[6] They found that Humanist Sans Serif typefaces with distinctly different letterforms were the most effective. The findings of the study inspired the typeface *Burlingame*, which was designed by Carl Crossgrove

The edible version of the typeface recreates the flavours of a long car journey to represent the inspiration behind *Burlingame's* design.

Coffee and mint liquorice-style caramels

Half the letters are created out of liquorice-style caramel for the tactile feel of a car's black leather or plastic dashboard. This is flavoured with coffee and mint for the 'stay awake' journey experience of service station takeaway coffees and mint sweets.

Zingy lemon jelly letters

In contrast, the illuminated numbers and letters are made from lemon jelly sweets. The zingy lemon contrasts with the deep coffee flavour and they shine brightly against the dark liquorice.

Serving suggestion: wrap the sweets individually and serve from pick 'n' mix bags for motorway service station authenticity. Ideal served up on long car journeys.

These sweets were originally created for Monotype. The recipe can be found on the Type Tasting blog.[7]

Gutenberg Bible courtesy of Trustees of Lambeth Palace Library

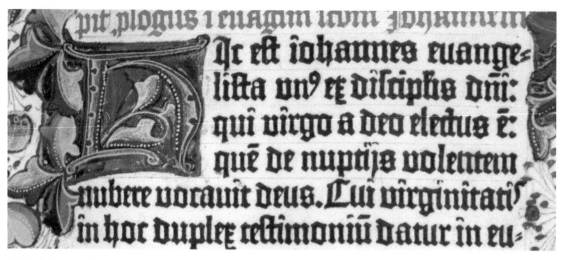

Edible history of type

Blackletter was the first printed type, created in Germany in the 1450s and based on the meticulous handwriting of scribes (see page 47).

I suggest this could be the hors d'oeuvres served at the beginning of a typographic banquet, which would take guests through the history of type with each course. It starts with the beginnings of print and the meal ends with a dessert inspired by fantastical ideas for the future of type made possible by developments in technology.

Gutenberg Bible hors d'oeuvres

An ornate illuminated letter made of spiced meat and pickled fruit arranged on a slice of rye bread. This is ancient bread that was a staple in Germany in the 1400s. 'Sliced bread' also functions as a metaphor for the mechanical reproduction made possible by the printing process.

The meat and fruits are to be laced with medieval spices and natural food colouring to bring the sculptural illuminated letter to life, and the pages of the book are printed in a Blackletter-style typeface with edible ink to look like a page from the *Gutenberg Bible* (above).

Eat your words

I often ask people what they think fonts would taste like. BBC Three Counties Radio presenter Nick Coffer thinks *Comic Sans* would be 'rocky road chocolate mousse with over-sickly marshmallows on the top and a bit of a dusting of icing sugar, perhaps with some glacé cherries.'[8]

London taxi driver Akin suggests *Times New Roman* would be a 'full English breakfast' because it is a filling version of an everyday meal, just like *The Times* newspaper.

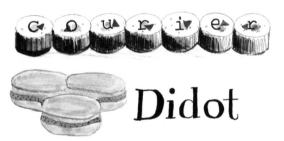

Birmingham City University students Edna Ma and Ian Nguyen proposed *Courier* to be a sushi roll, the equal spacing between each letter mirroring the monospaced typeface. Fellow student David Hansel suggested *Didot* would taste like classic French macaroons with a raspberry and butter-cream filling.

What typeface would you cook up and what would it taste like?

Cooking up a batch of *Futura*

Doughnut alphabet by Karina Monger

Processed cheese extra bold on rye

Chilli and soy sauce calligraphy by Alice Mazzilli

References

1 'What does type taste like?' by Sarah Hyndman, 2014, typetastingnews.com.

2 'Type on the tongue', *Eye* magazine blog.

3 '*Helvetica* water biscuits recipe' by Sarah Hyndman, 2014, typetastingnews.com.

4 '*Baskerville* Earl Grey tea biscuits recipe' by Sarah Hyndman, 2014, typetastingnews.com.

5 'What would Monotype's *Burlingame* typeface taste like?' by Sarah Hyndman, 2014, typetastingnews.com.

6 'Monotype Introduces the *Burlingame* Typeface Family', press release by Monotype, 2014, monotype.com.

7 Ibid. Sarah Hyndman.

8 'Talking "fonts"' on BBC Three Counties Radio, 2014, typetastingnews.com.

Online Type Tasting surveys can be found at typetasting.com.

sight   smell   touch   sound   taste

Eat Me — Decadent

EAT ME — economy

Eat Me — Instant

Eat Me — Diet food

INTUITIVE, REASONING, REFLECTIVE, FACULTIES.

LITERARY, OBSERVING, KNOWING, FACULTIES.

PHRENOL
BY
L.N. FOWL

Appendices

Props from Type Tasting events

Tested on Humans: Type Tasting experiments and surveys

Much research into typefaces explores legibility and focuses on the mechanics of letter forms and how they function. I am interested in how we respond to them emotionally. To find out more about this it has been important for me to talk to a wide range of people. As well as organising events, inventing games and asking lots of questions, I have created an ongoing series of 'Tested on Humans' surveys and experiments.

Initially these began as a way to test out the assumptions I make as a graphic designer, such as whether everybody recognises similar personalities in typefaces? To do this I particularly wanted to find out from non-designers how they respond to typefaces in their role as type consumers.

I am interested in using these experiments as a fun way to start conversations and to dispel the preconception that typography is a 'dry' subject for academics and experts. Typography is expressive; it is what our voices look like and it is how we express language visually. It reflects our everyday lives from highbrow to popular culture and the way we talk about it can reflect this cultural range.

Type Tasting surveys are intended both to gather information and to start conversations. Most are accessed via the Type Tasting website, although some are run from other places where they have no apparent links to typography.

The surveys collect a great deal of data, but they are not run under strictly scientific testing conditions. The intention is to highlight potential areas for future study in collaboration with scientists. Studies are currently being undertaken with some of the ideas initiated in the surveys.

Font Census (pages 80–85)
What personality types are different fonts? Participants are randomly assigned a font to answer questions about.
Each font has received 250–350 responses to date.

Emotions (page 66, 71)
What emotions do we see reflected in different letterforms and typefaces? This is an entry level survey aimed at non-designers.
Nearly 2,000 participants to date.

Eat Me (pages 104, 108, 110, 114, 116–118)
A range of surveys exploring what letterforms would taste like as food, and which investigate the visual codes of type on packaging design.
Various surveys both online and at events.

Font Fortunes (pages 4–5, 88–89, 92–93)
Are your values and aesthetics mirrored in the fonts you choose? Participants select from a range of options and receive a fun personality analysis based on the typeface they choose. They are asked how accurate they think the analysis is for their personality.
This demonstration is run at Type Tasting events.

Jellybean Demonstration (pages 114, 116, 117–118)
Can looking at different typefaces alter the taste of the jellybeans?
This demonstration is run at Type Tasting events.

Mood (pages 66–69, 71)
Various surveys include questions about the mood created by different typefaces and the results are cross-referenced.
Various surveys.

What Type of Pub? (pages 18, 53, 133)
This survey matches pubs to fonts and asks what type of establishment you think each would be from the style of the typeface.
Over 300 participants to date.

Sounds Like? (pages 39–41)
These surveys explore what typefaces would sound like. Some involve wearing headphones and matching sounds to fonts.
Over 500 participants to date.

Still or Sparkling? (page 105)
Which typefaces appear most sparkling or energetic, and which are more still or calming? The aim of this survey was to identify the features in a typeface that energise or calm it.
Over 500 participants to date.

Bitter/Sweet (page 104)
This is an ongoing project exploring whether different tastes can be represented as specific typefaces.
170 participants to date.

Type Dating Game (pages 93–99)
Participants are asked to choose a typeface to represent themselves for an evening at a speed dating event. They then select the type they would date, ditch and just be friends with. This explores whether we see personalities in typefaces that reflect our relationships with the people around us.
Over 5,000 participants to date.

Font Census breakdown of three typeface personality types in detail, from page 81

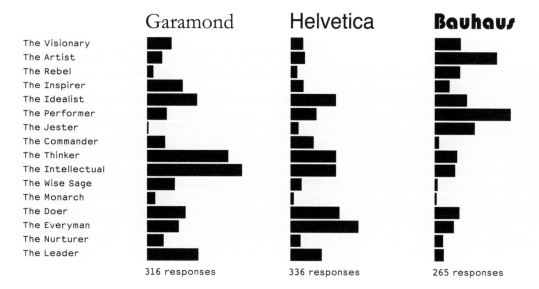

The Visionary
The Artist
The Rebel
The Inspirer
The Idealist
The Performer
The Jester
The Commander
The Thinker
The Intellectual
The Wise Sage
The Monarch
The Doer
The Everyman
The Nurturer
The Leader

Garamond — 316 responses

Helvetica — 336 responses

Bauhaus — 265 responses

Type Tasting online survey results. Participants are asked to describe what type of pub they think each is from the style of the typeface. Also see the 'Can you judge a pub by its sign' quiz on page 53.

 **The Ampersand Arms**

Didot: noisy, posh, wine bar, after work drinks, coffee and alcohol, sophisticated, above-average income, City workers.

THE AMPERSAND ARMS

Copesetic: art deco, cheeky, expensive, cocktails, bistro, women, gin, glitzy, wine bar, prohibition, small glasses, big prices.

The Ampersand Arms

Engravers' Old English: ye olde English pub, proper boozer, long-established, dimly lit, beer, ale, traditional, tavern.

THE AMPERSAND ARMS

Copperplate: country pub, open fire, traditional, pub cat, expensive, warm, touristy, gentrified, next to a law court.

 *The Ampersand Arms*

Flemish Script: fancy, dated, posh, a bit pretentious, middle-aged, brandy, expensive, older crowd, champagne.

 THE AMPERSAND ARMS

Trajan: upmarket, overpriced, wine bar, refined, civilised, traditional, stuffy, polite, decent, formal, good food, chain pub.

THE AMPERSAND ARMS

Buffalo Gal: rustic, saloon, new Western, country music, tattoos, burgers, whisky, dive bar, denim, good for brunch.

The Ampersand Arms

Amelia: older people, theme bar, Beatles' tribute bar, funky, fruity cocktails, cheap, kooky, hippy, classic rock, '60s bar.

 **The Ampersand Arms**

Cooper Black: seaside resort, retro, bargain boozer, warm beer, arcade machines, cheap, pre-drinking venue.

The Ampersand Arms

Times New Roman: pub for intellectuals, bland, boring, after work crowd, chain pub, gin and tonics, not expensive.

The Ampersand Arms

Braggadocio: inner city, casual, retro, young, hotel bar, cool, cocktails, airport, quirky, bright colours.

THE AMPERSAND ARMS

Mesquite: peanuts, tequila, hazy saloon, stuffed animals, whisky, bikers and old men, Americana, spirits, country music.

The Ampersand Arms

Helvetica: cheap, minimal interior, basic and bland, community, moody bar staff, cheap decor, metal furniture, cash only.

 **The Ampersand Arms**

Fette Fraktur: heavy metal music, seaside pub, all beards, pie, ale, small town, bitter, German beers, dark and smoky.

The Ampersand Arms

Comic Sans: juice bar for kids, coffee, cheap food, backpacker hostel, this wouldn't exist, basic drinks, uncool.

THE AMPERSAND ARMS

Caslon Titling: meetings, town, lawyers, doctors, expensive, a hidden drunk, wear a suit, sophisticated, classical music.

 THE AMPERSAND ARMS

Onyx: gastro-pub, high-class, pretentious, loud, cocktails, dancing, expensive club, smoking area, quirky, attractive exterior.

 THE AMPERSAND ARMS

Stencil: hole-in-the-wall dive, outskirts of a town, cheap, student, drink all you can, grunge music, happy hour, cheap shots.

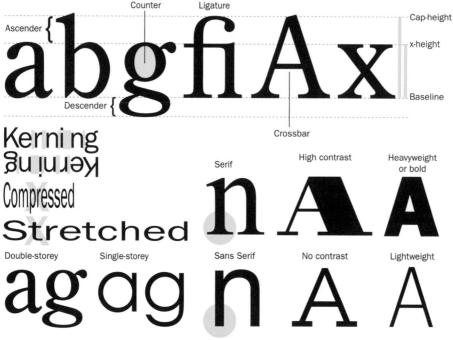

Display type
Headline
Body copy
Lorem ipsum dolor sit amet, consectetuer adipiscing elit, sed diam nonummy nibh euismod tincidunt ut laoreet dolore magna aliquam erat volutpat. Ut wisi enim ad minim veniam, quis nostrud exerci tation ullamcorper suscipit lobortis nisl ut aliquip ex ea commodo consequat.

Ranged left
Lorem ipsum dolor sit amet, consectetuer adipiscing elit, sed diam nonummy nibh euismod tincidunt ut laoreet dolore magna aliquam erat volutpat. Ut wisi enim ad minim veniam, quis nostrud exerci tation ullamcorper suscipit lobortis nisl ut aliquip ex ea commodo consequat.

Ranged right
Lorem ipsum dolor sit amet, consectetuer adipiscing elit, sed diam nonummy nibh euismod tincidunt ut laoreet dolore magna aliquam erat volutpat. Ut wisi enim ad minim veniam, quis nostrud exerci tation ullamcorper suscipit lobortis nisl ut aliquip ex ea commodo consequat.

Centred
Lorem ipsum dolor sit amet, consectetuer adipiscing elit, sed diam nonummy nibh euismod tincidunt ut laoreet dolore magna aliquam erat volutpat. Ut wisi enim ad minim veniam, quis nostrud exerci tation ullamcorper suscipit lobortis nisl ut aliquip ex ea commodo

Justified
Lorem ipsum dolor sit amet, consectetuer adipiscing elit, sed diam nonummy nibh euismod tincidunt ut laoreet dolore magna aliquam erat volutpat. Ut wisi enim ad minim veniam, quis nostrud exerci tation ullamcorper suscipit lobortis nisl ut aliquip ex ea commodo consequat. Duis

Ascender
Extends above the x-height.

Body copy
The main section of text to be read in a printed or online document (see above right).

Calligraphic
Influenced by handwriting.

Cap-height
The height of the capital letters.

Centred (see above right).

Contrast
The contrast between the thick and thin elements of type.

Counter
The space enclosed within a letter.

Crossbar
Horizontal stroke.

Descender
This extends below the baseline.

Display type
For large headlines (see above right).

Double-storey (see above).

Font
Traditionally a particular size, weight and style of a typeface (see page 27).

Glyph
An individual letter, character or symbol.

Italic
A cursive alphabet that slopes to the right.

Justified (see above right).

Kerning
The spacing between pairs of letters.

Leading
The spacing between the lines of text (separated by pieces of lead in traditional printing).

Ligature
Two or more letters combined to create a single glyph.

Monospaced
The width of all letters is the same and they are spaced out equally.

Pixels
Unit of measurement on screen.

Point (size). Unit of measurement in print.

Ranged left (see above right).

Ranged right (see above right).

Sans Serif
Without serifs.

Serif
With serifs.

Single-storey (see above).

Type
All characters or letters that are printed or shown on a screen.

Typeface
A particular design of type (page 27).

Typography
The art of arranging type.

Weight
How light or bold a font is.

x-height
The height of the lowercase letters. Note the curved letters will sit slightly outside.

Type classification* (See note over the page)

(See note over the page)

BLACKLETTER >>

Old English　　Fraktur　　Medieval 'Fairy tale'

SERIF >>

Roman inscriptions　　Humanist (Venetian)　　Old Style (Garalde)　　Transitional

Modern　　Fat Face　　Condensed　　Decorative

Slab Serif　　Typewriter　　Decorative

Engraved　　Glyphic (Flared)　　Tuscan　　Decorative

SANS SERIF >>

Grotesque　　Humanist　　Neo-Grotesque

SCRIPT >>

Geometric　　Bauhaus　　Art Deco　　De Stijl　　Formal　　Casual　　Art Nouveau　　Painted

■ Major categories
▪ Subsidiary styles

Blackletter
The first printed styles, which emulated the handwriting of monks and scribes.

Humanist (Venetian) Serif
The first printed Roman typefaces, inspired by the handwriting of Italian Renaissance scholars.

Old Style (Garalde) Serif
More refined letterforms, many of which are still popular today.

Transitional Serif
These have a more robust vertical axis, sharper serifs and show the transition from Old Style to Modern.

Modern (Didone) Serif
Improvements in technology and paper quality made it possible to print the extremely fine details in these typefaces.

Slab Serif
Typefaces with heavy serifs that were often printed at scale and became popular for headlines, billboards and posters.

Engraved
Inspired by engraved lettering.

Glyphic (Flared)
Often seen carved in stone, the flared shapes of these letter are created naturally by the flow of the signwriter's brush.

Grotesque Sans Serif
These are the early Sans Serifs that look a little like Serif typefaces but with the serifs removed.

Humanist Sans Serif
These have a hand-drawn feel to their shapes. If you try to draw a Sans Serif letter in freehand it will most likely reflect the flowing contours of Humanist curves.

Neo-Grotesque Sans Serif
These were the first Sans Serif typefaces to become commercially popular and are associated with the Swiss Modernist movement and with minimalism.

Geometric Sans Serif
These are based on geometric forms. They are sometimes considered to be less readable because the letter shapes look so similar.

Formal Script
This style of script is based on the handwriting of master calligraphers such as George Bickham and George Shelley.

Casual Script
These are informal letterforms that appear to have been written quickly. It is a popular signwriting style in the US.

References

* From page 135. Many design scholars have published various techniques of classification of typography. Most notably, in 1954 Maximilien Vox introduced a classification of type, which was later adopted by the Association Typographique Internationale (ATypI)[1]. This rigid classification system was designed prior to the mass computer-generated type design age and therefore did not allow for the evolution and variation of digitally designed typefaces we have today. However it has been a beneficial tool to review type design evolution in a chronological order. For the purpose of this publication I have simplified the classification to 'Major categories' and 'Subsidiary styles'.

1 This would later be set as British Standards Classification of Typefaces (BS 2961:1967).

Typefaces

Sources

Photographs

All photos and illustrations are by Sarah Hyndman,
unless credited otherwise.

Find out more

Stories about fonts/typefaces

Design Museum: Fifty Typefaces That Changed the World by John L. Walters (Conran).

Just My Type: A Book About Fonts by Simon Garfield (Profile Books).

Type: The Secret History of Letters by Simon Loxley (I.B. Tauris).

Creative lettering and type

The 3D Type Book by FL@33, Tomi Vollauschek and Agathe Jacquillat (Laurence King).

Typography Sketchbooks by Stephen Heller & Lita Talarico (Thames and Hudson).

Typoholic: Material Types in Design by Viction Workshop (Victionary).

Lettering and type in the environment

The Field Guide to Typography, Typefaces in the Urban Landscape by Peter Dawson (Thames & Hudson).

Signs: Lettering in the Environment by Phil Baines & Catherine Dixon (Laurence King Publishing).

Signpainters by Faith Levine (Princeton Architectural Press).

Typography

The Anatomy of Type: A Graphic Guide to 100 Typefaces by Stephen Coles (Harper Design).

The Geometry of Type by Stephen Coles (Thames & Hudson).

Stop Stealing Sheep and find out how type works by Erik Spiekermann & E.M. Ginger (Adobe Press).

Thinking with Type by Ellen Lupton (Princeton Architectural Press).

Type and Typography by Phil Baines & Andrew Haslam (Watson-Guptill).

Typography by Gavin Ambrose & Paul Harris (AVA Publishing).

Advanced typography

The Elements of Typographic Style by Robert Bringhurst (Hartley & Marks Publishers).

An Essay on Typography by Eric Gill (Penguin).

Crossmodal

The Perfect Meal: The multisensory science of food and dining by Charles Spence and Betina Piqueras-Fiszman (Wiley Blackwell).

Websites

Eye (International Review of Graphic Design)
eyemagazine.com

Ghostsigns archive
ghostsigns.co.uk

Grafik
grafik.net

I Love Typography
ilovetypography.com

It's Nice That
itsnicethat.com

Typographica
typographica.org

Watch Sarah Hyndman's TEDx talk Wake Up and Smell the Fonts:
bit.ly/wakeupandsmellthefonts

Mentioned

BBC Radio 4
bbc.co.uk/radio4

Crossmodal Research Laboratory
psy.ox.ac.uk/research/cross-modal-research-laboratory

D&AD
dandad.org

London College of Communication (University of the Arts)
arts.ac.uk/lcc

London Design Festival
londondesignfestival.com

Pick Me Up
Somerset House
somersethouse.org.uk

South by Southwest
sxsw.com

St Bride Library
14 Bride Lane, London EC4Y 8EQ
www.sbf.org.uk/library

TED/TEDx
ted.com

Victoria and Albert Museum
vam.ac.uk

With Relish
withrelish.co.uk

Thank you

Cheerleaders Theo S, Miho A, Syd H, Zoë C, the Paragon gang, Luke G, Alexandra B, Alex F, Claire M, Hazel G, Becky C, Caryl J, Gaynor M, Natalie C, Helen R, Andy J, Minna O, Sarah L-K, Beverley G, Nicky M and Rona S.

Emily Gosling, Angus Montgomery, Emma Tucker and John L. Walters for taking me seriously. The Monotype team for taking me 'just seriously enough'.

Eagle-eyed Caryl Jones, Vinita Nawathe, Theo Stewart, Geraldine Marshall and Eugenie Smit.

Volunteers Eunjung Ahn, Miho Aishima, Lissy Bonness, Emily Bornoff, Zoë Chan, Paul Fine, Eva Gabor, Syd Hausmann, Chelsea Herbert, Stuart Hollands, Natalie Kelter, Lucy McArthur, Martin Naidu, David Owens, Lucy Pendlebury, Eugenie Smit, Peter Strauli, Qian Yuan and Nicola Yuen

Live letterers extraordinaire Oli Frape and Ruth Rowland. Everybody who took part in the exhibition and workshops at the V&A.

Makers of edible type, especially Alice Mazzilli, Andreja Brulc, Christine Binns, Julie Mauro, Karina Monger and Stephen Boss.

The knowledgeable signpainting crew led by Sam Roberts, Ash Bishop, the inimitable Mike Meyer and his Brushettes.

Professor Charles Spence, Andy Woods and Carlos Velasco Pinzon.

Everybody who has taken part in the surveys, experiments and live events.

To future type explorers Leon and Stefan, enjoy the adventure.

Written and designed by Sarah Hyndman
Cover designed by Two Associates
Proofread by Caroline McArthur

Read the personality analysis that matches the Font Fortunes typeface you chose at the start of the book.

Baskerville

Choice #1

Helvetica

Choice #2

Didot

Choice #3

You have traditional values and can be a bit of a perfectionist at times.

You prefer an intellectual or academic conversation over a display of emotions. Your style is of quiet refinement and you are the opposite of ostentatious. You are seen as wise and a little conventional, but you are forward-thinking and not old-fashioned.

You read a great deal and enjoy a well-made cup of tea.

Your friends trust you to give well-informed and considered opinions, and that if you don't know the answer you'll just say so (rather than making something up).

Secret guilty pleasure: enjoying the smell of quality when you unwrap something new like a wallet, handbag or a freshly printed book.

Baskerville is a Transitional Serif typeface which was designed by John Baskerville in Birmingham, England in the 1750s.

You are practical and versatile, and able to turn your hand to many things. You have a pragmatic and calm approach, which may give the impression that you are conventional and not a risk-taker.

You like to take a neutral viewpoint and weigh up a situation before voicing a well-considered opinion. You will often find yourself as the mediator in a heated situation.

You are a minimalist at heart and feel happiest living in an uncluttered environment, ideally with views that stretch all the way to the horizon.

Secret guilty pleasure: dressing up for a stylish event, or an outlandish fancy dress party. You will give the impression that you need to be coerced into this, but secretly you really enjoy it.

Helvetica is a neo-Grotesque Sans Serif. It was created by Max Miedinger and colleagues in 1957.

You come across as poised and thoughtful and this sometimes gives you an air of importance. However, this outward appearance of confidence hides your more reserved personality. You are a people watcher but, rather than doing this from the shadows, you are happy to stand in the middle of the room and to be seen.

You always know exactly how to dress for a situation, whether it is the glamour of a party or a casual weekend away.

You think it is the small, finishing touches that reveal the most about a person.

Secret guilty pleasure: doing something indulgent and messy like eating chocolate in the bath or wearing flip flops in the mud.

Didot is a Modern (or Didone) typeface. It is based on typefaces created by Firmin and Pierre Didot in Paris in the late 18th and early 19th centuries.

Gill Sans

Choice #4

Clarendon

Choice #5

You are a traditionalist who likes to think of yourself as being modern.

You are articulate and you use the correct spelling and grammar in emails, texts and tweets — no abbreviations or acronyms from you. But this does not mean you are overly formal, far from it in fact. Your friends see you as an everyman (or woman), a regular Joe, one of the gang.

You have a cheerful, outgoing demeanour with a slightly idiosyncratic manner (which can occasionally frustrate those closest to you).

You are practical and you always know what to do in a situation, so you will often be the person people turn to when they have a problem or need advice.

Secret guilty pleasure: swearing.

Gill Sans is a Humanist Sans Serif typeface designed by Eric Gill in England in the 1920s.

You have that quiet, assured confidence that means you do not have to shout to get attention. You are a natural leader although you do not always choose (or want) to be.

You have both feet firmly on the ground and have a love of the outdoors. You enjoy wide, open spaces and the freedom that comes from walking in the countryside.

You are a practical person and if there's a problem you would prefer to work out how to solve it yourself rather than delegating the task to others.

You are more of a doer than a thinker and all your views are your own.

Secret guilty pleasure: enjoying a nostalgic moment like crying at a sentimental film in the dark.

Clarendon is a Slab Serif typeface that was created by Robert Besley in England in the 1850s.

Typefaces used in the book

The cover and body copy in this book are in **Franklin Gothic**, which I chose because it is a readable classic. Styles like this were used for early American tabloid newspaper headlines and this populist association reminds me not to take myself too seriously.

The large titles are in **Balega**, which is a bold, charismatic display face that looks interesting and feels a little challenging.

I chose Monotype Century Expanded for the references and picture captions because it is directing you to more in-depth information and feels knowledgeable.

Courier Sans is used for instructions and data because its open letters are clear and easy to read.

Franklin Gothic by Morris Fuller Benton, 1902.

Balega by Jürgen Weltin, 2003.

Monotype Century Expanded by Linn Boyd Benton and T. L. DeVinne, 1894.

Courier Sans by James Goggin, 1994–2001.